# Fauna of New Zealand
## Ko te Aitanga Pepeke o Aotearoa

## INVERTEBRATE SYSTEMATICS ADVISORY GROUP

### REPRESENTATIVES OF LANDCARE RESEARCH
**Dr D.R. Penman**
*Landcare Research
Lincoln Agriculture & Science Centre
P.O. Box 69, Lincoln, New Zealand*

**Dr T.K. Crosby** and **Dr M.-C. Larivière**
*Landcare Research
Mount Albert Research Centre
Private Bag 92170, Auckland, New Zealand*

### REPRESENTATIVE OF UNIVERSITIES
**Dr R.M. Emberson**
*Ecology and Entomology Group
Soil, Plant, and Ecological Sciences Division
P.O. Box 84, Lincoln University, New Zealand*

### REPRESENTATIVE OF MUSEUMS
**Mr R.L. Palma**
*Natural Environment Department
Museum of New Zealand Te Papa Tongarewa
P.O. Box 467, Wellington, New Zealand*

### REPRESENTATIVE OF OVERSEAS INSTITUTIONS
**Dr J.F. Lawrence**
*CSIRO Division of Entomology
G.P.O. Box 1700, Canberra City
A.C.T. 2601, Australia*

\* \* \*

### SERIES EDITOR
**Dr T. K. Crosby**
*Landcare Research
Mount Albert Research Centre
Private Bag 92170, Auckland, New Zealand*

Fauna of New Zealand
Ko te Aitanga Pepeke o Aotearoa

Number / Nama 40

# Cixiidae
(Insecta: Hemiptera: Auchenorrhyncha)

M.-C. Larivière

Landcare Research
Private Bag 92170, Auckland, New Zealand
LariviereM@landcare.cri.nz

Manaaki
Whenua
P R E S S

Lincoln, Canterbury, New Zealand
1999

Land Invertebrates". Nāna taua kaupapa i whakahaere i te tau 1994 ki te tau 1997. I ēnei rā, he tūranga kairangahau ukiuki tōna i Manaaki Whenua, he whakarōpū koiora te kaupapa matua. Arā anō tētahi o āna mahi ko te tiaki i ngā Coleorrhyncha, ngā Auchenorryncha me ngā Heteroptera kei roto i te Kohinga Angawaho o Aotearoa. Neke atu i te 60 ngā tuhinga kua oti i a Marie-Claude mō te whakapapa o ētahi whānau o ngā aitanga a Punga, tae atu ki ngā momo ka kitea i tēnā wāhi, i tēnā wāhi, me te āhua o tā rātou noho. Kua puta i a ia he tuhinga e pā ana ki te Orthoptera, kua whai wāhi atu hoki ki ētahi tānga kōrero e pā ana ki te Carabidae (Coleoptera). Tērā anō tētahi tino kaupapa e whāia ana e Marie-Claude, ko te pārongo-koiora, tae atu ki te whakarōpū ā-mati, te hanga whakaahua ki te rorohiko, me te pānui kōrero ki te pae tukutuku.

**He Kupu Āwhina — Glossary**
ahi tipua — volcano
ahurei — unique, uniqueness
Kawatiri — Buller River
kukune — evolve
manoao — barrier pine
mōriroriro — isolated, remote
papatahi — flattened
pārongo-koiora — bioinformatics
pīakaaka — roots
pūhihi — antenna
puninga — genus
raurarahi — broadened
tawhai — beech family
Te Taka o te Kārahu a Tamatea — Fiordland
Te Uri Māroa — Gondwanaland
tōtorowhiti — grass tree
whakarōpū ā-mati — digital taxonomy

Translation by **H. Jacob**
Huatau Consultants, Levin

## DEDICATION

*I dedicate this work to Vernon R. Vickery, whose wisdom, friendship, and guidance have been a steady source of support during my undergraduate and graduate studies, and through the development of my career until this day. Dr Vickery's enthusiasm as a field entomologist and his numerous publications on the faunistics of Nearctic orthopteroids have been inspirational to my research in Canada as well as here in New Zealand. Vick's extreme generosity and loyalty towards his students, his open-mindedness towards their study goals and scientific aspirations, his integrity towards their achievements, and his discreet yet attentive supervision style, make him a wonderful mentor.*

*Marie-Claude*

Dr V.R. Vickery spent the major part of his professional career at the Macdonald Campus of McGill University (Ste-Anne-de-Bellevue, Québec, Canada) where he taught entomology to several generations of students who dispersed throughout the world to lead their entomological careers.

Dr Vickery is widely known and respected for his research on Nearctic orthopteroids. As Curator of the Lyman Entomological Museum and Research Laboratory, he was largely responsible for developing the collections as a nationally significant resource on the insects of Canada. He made a very important contribution as Editor of the Museum's *Memoirs* for several years, and he contributed substantially to the development of the Orthopterist Society.

Vernon R. Vickery is now enjoying retirement in his native Nova Scotia where he started his entomological career.

## ABSTRACT

The New Zealand Cixiidae fauna (11 genera, 25 species) is reviewed. Two new genera are described: *Chathamaka* gen. nov. and *Parasemo* gen. nov. Three new synonymies are established: *Cixius interior* Walker = *Cixius punctimargo* Walker, *Malpha iris* Myers = *Malpha muiri* Myers, *Semo westlandiae* Larivière & Hoch = *Huttia harrisi* Myers. *Cixius interior* Walker is removed from the genus *Koroana* Myers, and *Cixius rufifrons* Walker is resurrected from synonymy and reinstated as the type species of *Koroana*. *Huttia harrisi* Myers is transferred to the genus *Semo* White. The name *Cixius aspilus* Walker is found to have been established on a hybrid. Eight new species are described: *Aka dunedinensis* sp. nov., *A. rhodei* sp. nov., *A. westlandica* sp. nov., *Chathamaka andrei* sp. nov., *Cixius inexspectatus* sp. nov. (previously recorded as *C. aspilus*), *C. triregius* sp. nov., *Huttia northlandica* sp. nov., and *Parasemo hutchesoni* sp. nov. A concise review of the taxonomy of all taxa is presented. Genera and species are keyed. Concise descriptions are provided together with illustrations emphasising the most important diagnostic features of the external morphology and male genitalia. Information is given on synonymy, type data, material examined, geographical distribution, and biology.

Keywords. Hemiptera, Auchenorrhyncha, Cixiidae, new genera, new species, taxonomy, keys, distribution, ecology, fauna.

Larivière, M.-C. 1999: Cixiidae (Insecta: Hemiptera: Auchenorrhyncha). *Fauna of New Zealand 40*, 93 pages.

*Received: 17 December 1998. Accepted: 21 April 1999.*

## CHECKLIST OF TAXA

Family **CIXIIDAE**
Subfamily **CIXIINAE**
Tribe **CIXIINI**

Genus *Aka* White, 1879 .......... 18
  *dunedinensis* sp. nov .......... 19
  *duniana* (Myers, 1924) .......... 19
  *finitima* (Walker, 1858) .......... 20
  *rhodei* sp. nov. .......... 21
  *westlandica* sp. nov. .......... 22

Genus *Chathamaka* gen. nov. .......... 23
  *andrei* sp. nov. .......... 23

Genus *Cixius* Latreille, 1804 .......... 24
  *inexspectatus* sp. nov. .......... 25
  *punctimargo* Walker, 1858 .......... 26
  *triregius* sp. nov. .......... 27

Genus *Confuga* Fennah, 1975 .......... 28
  *persephone* Fennah, 1975 .......... 28

Genus *Huttia* Myers, 1924 .......... 28
  *nigrifrons* Myers, 1924 .......... 29
  *northlandica* sp. nov. .......... 30

Genus *Koroana* Myers, 1924 .......... 30
  *arthuria* Myers, 1924 .......... 31

  *lanceloti* Larivière, 1997 .......... 32
  *rufifrons* (Walker, 1858) stat. nov. .......... 33

Genus *Malpha* Myers, 1924 .......... 35
  *cockcrofti* Myers, 1924 .......... 35
  *muiri* Myers, 1924 .......... 36

Genus *Parasemo* gen. nov. .......... 37
  *hutchesoni* sp. nov. .......... 37

Genus *Semo* White, 1879 .......... 37
  *clypeatus* White, 1879 .......... 39
  *harrisi* (Myers, 1924) comb. nov. .......... 40
  *southlandiae* Larivière & Hoch, 1998 .......... 41
  *transinsularis* Larivière & Hoch, 1998 .......... 41

Tribe **OECLEINI**

Genus *Tiriteana* Myers, 1924 .......... 43
  *clarkei* Myers, 1924 .......... 43

Tribe **PENTASTIRINI**

Genus *Oliarus* Stål, 1862 .......... 44
  *atkinsoni* Myers, 1924 .......... 44
  *oppositus* (Walker, 1851) .......... 45

In addition, *Cixius kermadecensis* Myers, 1924 (not treated here) has been recorded from the Kermadec Islands.

repository acronyms. A list of geographical coordinates for the main localities from which material was examined, is given in Appendix A.

Biological notes are based on an analysis of specimen label data. The native plants associated with New Zealand Cixiidae are listed in Appendix B, along with their family placement.

Very little specific or reliable information on distribution and biology could be found in the New Zealand literature (over 100 papers) which was thoroughly scanned.

The status, depositories, and full label data of all primary type specimens seen (and a summary of label information for secondary type specimens) are cited for each species. In the list of label data, different labels are separated by a solidus (/) and different lines on a label by a semicolon; all other punctuation is as it appears on the label.

**Collecting and preparation.** Adults of Cixiidae are generally collected by sweeping or beating shrubs and trees and, for some species, the low vegetation. Beating or handpicking from selected plants, especially host plants, yields invaluable biological information. Eggs and nymphs can be collected when the host plant is known by digging and sifting the soil around the roots of the host plant.

Adults are best preserved dry. Eggs and nymphs are usually kept in 70–75% ethanol. If a molecular study is intended, adults as well as immatures can be kept in 100% ethanol.

Preparation and curation of insects have been fully described by Walker and Crosby (1988). All specimens should be collected with the locality (including area code: Crosby *et al.* 1998, and geographical coordinates such as latitude and longitude), collection date, collector's name, and ecological data (e.g., general habitat and host plant).

Most features of the external morphology and the male genitalia can be viewed under an ordinary dissecting microscope. It is necessary to relax and dissect male specimens to study their genitalia.

Male genitalia can be dissected as follows. Pinned specimens are warmed for 5–10 minutes in soapy water or hot alcohol (70–75% ethanol). If the abdomen alone is to be used, this can be separated from the rest of the body by inserting a pin between it and the thorax, or, in difficult cases, by first relaxing the whole specimen in hot soapy water. Each specimen or abdomen is transferred to a watch glass half-filled with water (if soapy water was used) or ethanol, and the pygofer (genital capsule) is pulled away from the body using fine forceps and a micro-scalpel (needle tip from a 1.0 ml disposable hypodermic syringe). The pygofer is then warmed in very hot (almost boiling) ethanol for about 5 minutes, then transferred to another watch glass also containing ethanol. The anal tube, genital styles and aedeagus are detached and extracted from the pygofer in this solution, using fine forceps and a micro-scalpel. Dissected genitalia are subsequently stored in genitalia vials containing glycerine, and remounted on the pin below the relevant specimens.

Another method uses a 10% KOH solution to macerate the abdomen. This is done by putting the abdomen in a test tube containing a little KOH solution and by placing the test tube on a hot plate in a beaker containing water to warm up its content for 10–15 minutes (alternatively the abdomen can be left in cold KOH overnight). The rest of the dissection is done in water. This technique may be quicker for routine identification; KOH is a clearing agent that allows the examination of genitalia by transparency. If the purpose of the study is a taxonomic revision, it is preferable to extract the structures contained in the pygofer because they need to be viewed and illustrated in the exact same angle or position.

**Taxonomically relevant characters.** Detailed investigations of the external morphology, including the tymbal organ and the female abdomen, revealed very few characters that can be useful in species diagnosis. In most genera, species are morphologically similar, and intraspecific variation is high in most characters, e.g., forewing venation, shape and proportion of structures of the head and thorax.

The characters presented in the descriptions are subsets of the totality of adult characters studied, and represent the most important differences between, or variation amongst, closely related taxa. Characters or states of characters not mentioned in the species descriptions are as described in generic descriptions.

Descriptive measurements and counts were taken in the following manner: *vertex length* measured from tip of basal emargination to apex of vertex; *vertex width* taken at level of tip of basal emargination; *forewing length* taken from base to apex; *forewing width* measured at tip of clavus; *body length* measured from apex of head to tip of forewing, cited as a range with mean in parentheses.

Characters with the highest diagnostic value at the species level have been illustrated, including the most diagnostic aspects of the male genitalia. Most illustrations provided in this work represent the most commonly encountered state of a character. The user must allow some degree of variation when working with individual specimens, especially in the case of newly emerged adults that may still have a soft cuticle. It is not uncommon to see somewhat distorted genital structures in teneral individuals.

Keys are somewhat artificial. They are intended as an aid to identification, not a statement of the author's opin-

## ABSTRACT

The New Zealand Cixiidae fauna (11 genera, 25 species) is reviewed. Two new genera are described: *Chathamaka* gen. nov. and *Parasemo* gen. nov. Three new synonymies are established: *Cixius interior* Walker = *Cixius punctimargo* Walker, *Malpha iris* Myers = *Malpha muiri* Myers, *Semo westlandiae* Larivière & Hoch = *Huttia harrisi* Myers. *Cixius interior* Walker is removed from the genus *Koroana* Myers, and *Cixius rufifrons* Walker is resurrected from synonymy and reinstated as the type species of *Koroana*. *Huttia harrisi* Myers is transferred to the genus *Semo* White. The name *Cixius aspilus* Walker is found to have been established on a hybrid. Eight new species are described: *Aka dunedinensis* sp. nov., *A. rhodei* sp. nov., *A. westlandica* sp. nov., *Chathamaka andrei* sp. nov., *Cixius inexspectatus* sp. nov. (previously recorded as *C. aspilus*), *C. triregius* sp. nov., *Huttia northlandica* sp. nov., and *Parasemo hutchesoni* sp. nov. A concise review of the taxonomy of all taxa is presented. Genera and species are keyed. Concise descriptions are provided together with illustrations emphasising the most important diagnostic features of the external morphology and male genitalia. Information is given on synonymy, type data, material examined, geographical distribution, and biology.

Keywords. Hemiptera, Auchenorrhyncha, Cixiidae, new genera, new species, taxonomy, keys, distribution, ecology, fauna.

Larivière, M.-C. 1999: Cixiidae (Insecta: Hemiptera: Auchenorrhyncha). *Fauna of New Zealand 40*, 93 pages.

*Received: 17 December 1998. Accepted: 21 April 1999.*

## CHECKLIST OF TAXA

Family **CIXIIDAE**
Subfamily **CIXIINAE**
Tribe **CIXIINI**

Genus *Aka* White, 1879 ............................................ 18
   *dunedinensis* sp. nov. ........................................ 19
   *duniana* (Myers, 1924) ........................................ 19
   *finitima* (Walker, 1858) ...................................... 20
   *rhodei* sp. nov. ............................................... 21
   *westlandica* sp. nov. ......................................... 22

Genus *Chathamaka* gen. nov. ................................... 23
   *andrei* sp. nov. ............................................... 23

Genus *Cixius* Latreille, 1804 ................................... 24
   *inexspectatus* sp. nov. ....................................... 25
   *punctimargo* Walker, 1858 ................................... 26
   *triregius* sp. nov. ............................................ 27

Genus *Confuga* Fennah, 1975 .................................. 28
   *persephone* Fennah, 1975 ................................... 28

Genus *Huttia* Myers, 1924 ...................................... 28
   *nigrifrons* Myers, 1924 ...................................... 29
   *northlandica* sp. nov. ....................................... 30

Genus *Koroana* Myers, 1924 .................................... 30
   *arthuria* Myers, 1924 ........................................ 31

   *lanceloti* Larivière, 1997 .................................... 32
   *rufifrons* (Walker, 1858) stat. nov. ....................... 33

Genus *Malpha* Myers, 1924 ..................................... 35
   *cockcrofti* Myers, 1924 ...................................... 35
   *muiri* Myers, 1924 ............................................ 36

Genus *Parasemo* gen. nov. ...................................... 37
   *hutchesoni* sp. nov. .......................................... 37

Genus *Semo* White, 1879 ........................................ 37
   *clypeatus* White, 1879 ....................................... 39
   *harrisi* (Myers, 1924) comb. nov. ........................ 40
   *southlandiae* Larivière & Hoch, 1998 ................. 41
   *transinsularis* Larivière & Hoch, 1998 ................ 41

Tribe **OECLEINI**

Genus *Tiriteana* Myers, 1924 ................................... 43
   *clarkei* Myers, 1924 .......................................... 43

Tribe **PENTASTIRINI**

Genus *Oliarus* Stål, 1862 ....................................... 44
   *atkinsoni* Myers, 1924 ....................................... 44
   *oppositus* (Walker, 1851) ................................... 45

In addition, *Cixius kermadecensis* Myers, 1924 (not treated here) has been recorded from the Kermadec Islands.

## CONTENTS

Acknowledgments .................................................. 10
Introduction ........................................................ 10
Morphology and terminology ................................. 15
Methods and conventions ...................................... 15
Key to genera ...................................................... 17
The taxa of New Zealand Cixiidae ........................... 18
References .......................................................... 46
Appendix A: Geographical coordinates of main collection localities ................................................ 48
Appendix B: Native plants associated with Cixiidae species ................................................................ 52
Figures and Maps ................................................. 53
Taxonomic index .................................................. 87

## ACKNOWLEDGMENTS

For the opportunity to borrow and examine material in their care I thank R.M. Emberson and J.M.W. Marris (Lincoln University, Lincoln), P.M. Johns (University of Canterbury, Christchurch), M.W. Walker and S.D. Pollard (Canterbury Museum, Christchurch), M.J. Nuttall (Forest Research Institute, Rotorua), B.H. Patrick and A.C. Harris (Otago Museum, Dunedin), J.W. Early (Auckland Institute and Museum), R.L. Palma (Museum of New Zealand Te Papa Tongarewa, Wellington), M.J. Fletcher (NSW Agriculture, Orange), T.A. Weir (Australian National Insect Collection, Canberra), G.F. Gross (South Australian Museum, Adelaide), J. Margerison-Knight and M.D. Webb (Natural History Museum, London).

Thanks are extended to G. Hall, D. M. Gleeson, and L. H. Clunie (Landcare Research, Auckland) for additional material, invaluable biological information and/or technical assistance.

I express my gratitude to D.W. Helmore (Landcare Research, Auckland) for habitus and face illustrations of adults and Fig. 13–27 and to T.K. Crosby and R.E. Beever (Landcare Research, Auckland) for their encouragement and for allocating resources towards completion of this work.

Special thanks to B.E. Rhode (Landcare Research, Auckland) who has prepared most dissections of male genitalia, recorded specimen label data and worked extensively on the geographic database. Without her exceptional dedication the completion of this work would, without doubt, have been delayed for a number of years.

I am particularly grateful to G.H. Sherley (formerly Department of Conservation, Wellington) and Chris J. Green (Department of Conservation, Auckland) for their help in obtaining collecting permits without which much of the biological and distributional information presented here would have been unobtainable. I also wish to thank all other Department of Conservation staff who have provided field assistance and access to Conservation lands.

Julian H. Cone (Landcare Research, Lincoln) and D. J. Thoreau (Landcare Research, Auckland) have been very helpful in providing Geographic Information System (ArcView) support and other computer support. I also owe many thanks to Landcare Research (Lincoln) and HortResearch (Auckland) library staff for their assistance in obtaining important literature references for this work.

I am particularly grateful to M. J. Fletcher (NSW Agriculture, Orange) and A.F. Emeljanov (Russian Academy of Sciences, St. Petersburg, Russia) for their comments and help with the final manuscript.

Finally, I thank my husband, A. Larochelle, for his helpful comments on and criticisms of the manuscript, for many highly productive and enjoyable field ventures, and for his continued encouragement during this project.

This work has been supported by the Foundation for Research Science and Technology under Contract C09617.

## INTRODUCTION

The family Cixiidae is cosmopolitan, with over 1500 described species that may represent about 40 percent of the actual world fauna. The present faunal review records 11 genera and 25 species for New Zealand. This should constitute the near totality of the fauna; a few additional cave-dwelling species may still be discovered.

Compared with New Zealand, the Australian fauna remains largely unknown, with only about 40 species described, of which several are currently placed in the cosmopolitan genus *Oliarus* Stål.

The present work offers a concise faunal review of the New Zealand Cixiidae, based on the study of adults contained in local and overseas collections. It represents a first modern attempt to bring together the scattered information dealing with the group.

The goals of this review are straightforward: to provide an inventory of New Zealand taxa, a concise treatment of their taxonomy, keys to genera and species, and a summary of the available information on species distribution and biology.

**Taxonomic history.** There has been little work done on the faunistics of the New Zealand Cixiidae since the beginning of the 20th century. The most comprehensive taxonomic treatment to date is that of Myers (1924) who provided a key to eight genera, descriptions of new taxa (4 genera, 9 species), and comments on most of the 17 species known at the time. A limited amount of study material, mostly types, was available to Myers. As a result, the high degree of morphological variation between and within species may not have appeared as obvious as it is today from the extensive material contained in New Zealand collections. For this reason, it seemed timely to provide a new taxonomic revision for this group, one that includes descriptions and keys that take into account this new information.

Metcalf (1936) was the first to catalogue the New Zealand taxa following Myers' paper. It took about 40 years before another taxonomic paper was published on Cixiidae, i.e., the description of the cave-dwelling species *Confuga persephone* by Fennah (1975). Subsequently, Wise (1977) published a synonymical list of the New Zealand taxa. Deitz and Helmore (1979) provided an easy-to-follow key to identify Cixiidae from other planthopper families occurring in New Zealand and a key to Cixiidae genera. Larivière (1997a) updated the list of described taxa. Finally, Larivière (1997b) and Larivière & Hoch (1998) resumed work on the taxonomy of these planthoppers by reviewing *Koroana* Myers and *Semo* White.

**Higher classification.** The main contributors to the higher classification of the family include Muir, Metcalf, and Emeljanov. Muir (1922) erected the tribe Oecleini when he described the Malayan genus *Euryphlepsia*. His subsequent papers (1923, 1925) dividing the family into 2 tribes (Cixiini, including Oecleini, and Bothriocerini) became the basis for most work since. In his catalogue, Metcalf (1936) adopted the classification proposed by Muir (1925), but in 1938 he elevated the tribes Cixiini and Bothriocerini to the rank of subfamilies and created new tribes within each subfamily. His tribal divisions, however, have not been universally adopted by subsequent workers. Emeljanov (1971) separated the tribe Pentastirini, with two subtribes, from Cixiini. More recently, Emeljanov (1989) reviewed the problem of the higher classification of Cixiidae.

There is as yet no rigorous treatment of the classification for the family. Criteria defining supraspecific taxa are still insufficiently elaborated, and phylogenetic (or even cladistic) analyses are practically nonexistent for this family.

The family classification adopted here follows Emeljanov (1989) who recognised three subfamilies (Bothriocerinae, Borystheninae, and Cixiinae). Three tribes of Cixiinae have so far been recorded from New Zealand: the Oecleini, represented by *Tiriteana* [and possibly *Confuga* (see Fennah 1975: 379)], the Pentastirini, represented by *Oliarus*, and the Cixiini, represented by all other genera.

Emeljanov (1997) presented a tentative cladogram of Cixiidae tribes, in an attempt to evaluate the usefulness of certain characters to reconstruct phylogeny. He suggested that *Semo* could be placed in a separate tribe, the Semonini, together with the mainly Oriental genera *Kuvera* and *Betacixius*, on the basis of having a swollen postclypeus. Emeljanov made it clear, however, that his attempt to distinguish the tribes of Cixiidae was made on precarious bases; he did not intend to propose a formal classification. Therefore, the present revision follows Larivière & Hoch (1998) in adopting the traditional placement of *Semo* within Cixiini.

**Morphological characteristics.** Cixiidae have retained a series of plesiomorphic characters distinguishing them from other fulgoroid families, for example, the presence of a third ocellus on the frons in many genera. For this reason they are considered to be one of the most primitive families of planthoppers.

They are characterised mainly by the forewings, which are usually membranous, and usually have tubercles bearing setae (setiferous tubercles) along the veins. Their size generally ranges from 3 to 11 mm and most species hold their wings horizontally, although some hold them vertically. Males have complex genital structures that are partially exposed. The aedeagus is composed of a shaft (periandrium) and a flagellum with a number of processes and appendages. These structures are usually highly diagnostic at the species level. Females have a sword-like or short porrect ovipositor. Figures 1–11 provide a basic understanding of the morphological structures used to identify cixiid genera and species.

**Geographical distribution.** The New Zealand fauna is highly insular, with 8 genera (73%) and 25 species (100%) presently recorded as endemic. The indigenous genus *Aka* White has two representatives in Tasmania. The genera *Oliarus* and *Cixius* are cosmopolitan.

The overall distribution of New Zealand Cixiidae is summarised in Table 1 and in Maps 13 and 14. Species distributions are clearly defined and largely allopatric. Four genera (*Cixius*, 3 species; *Huttia* Myers, 2 species; *Parasemo* gen. nov., 1 species; *Tiriteana* Myers, 1 species) are confined to the North Island. The cave-dwelling genus *Confuga* (1 species) is only known from the northwest of the South Island. *Chathamaka* gen. nov. (1 species) is endemic to the Chatham Islands.

The great majority of Cixiidae species occur on the North Island, although they are not all restricted to it. Seven species (*Aka duniana* (Myers), *Koroana rufifrons* (Walker),

**Table 1.** Summarised distributions for New Zealand Cixiidae. Two-letter area codes follow Crosby *et al.* (1998)

| Species | North Island | | | | | | | | | | | |
|---|---|---|---|---|---|---|---|---|---|---|---|---|
| | ND | AK | CL | WO | BP | TK | TO | HB | GB | RI | WI | WN | WA |
| *Aka dunedinensis* | | | | | | | | | | | | | |
| *A. duniana* | | | | | | | TO | HB | | | | WN | |
| *A. finitima* | ND | AK | CL | | BP | | TO | | | RI | WI | WN | |
| *A. rhodei* | | | | | | | TO | | | RI | | | |
| *A. westlandica* | | | | | | | | | | | | | |
| *Chathamaka andrei* | | | | | | | | | | | | | |
| *Cixius inexspectatus* | ND | AK | CL | | | | TO | | | | | | |
| *C. punctimargo* | ND | AK | CL | | BP | | | | GB | | | | |
| *C. triregius* | | | | | | | | | | | | | |
| *Confuga persephone* | | | | | | | | | | | | | |
| *Huttia nigrifrons* | ND | AK | CL | | BP | | TO | | | | | WN | |
| *H. northlandica* | ND | | | | | | | | | | | | |
| *Koroana arthuria* | | | | | | | | | | | | | |
| *K. lanceloti* | | | | | | | | | | | | | |
| *K. rufifrons* | ND | AK | CL | WO | BP | TK | TO | HB | GB | RI | WI | WN | WA |
| *Malpha cockrofti* | | | | | | | | | | | | | |
| *M. muiri* | | | | | | | | | | | | WN | |
| *Oliarus atkinsoni* | ND | AK | CL | | BP | TK | TO | | | | | WN | |
| *O. oppositus* | ND | AK | CL | WO | BP | TK | TO | HB | GB | RI | WI | WN | WA |
| *Parasemo hutchesoni* | | | | | BP | | TO | | | | | | |
| *Semo clypeatus* | | | | | | TK | TO | | GB | | | | |
| *S. harrisi* | | | | | | | | | | | | | |
| *S. southlandiae* | | | | | | | | | | | | | |
| *S. transinsularis* | | | | | | | TO | | | RI | | WN | |
| *Tiriteana clarkei* | ND | AK | CL | WO | BP | TK | TO | | GB | | | WN | |

*Malpha muiri* Myers, *Semo clypeatus* White, *S. transinsularis* Larivière & Hoch, *Oliarus atkinsoni* Myers, and *O. oppositus* (Walker)) also occur on the South Island, mostly in northernmost areas. *Oliarus oppositus* is widely distributed on both islands.

Seven species (*Aka westlandica* sp. nov., *A. dunedinensis* sp. nov., *Koroana arthuria* Myers, *K. lanceloti* Larivière, *Malpha cockcrofti* Myers, *Semo harrisi* (Myers), and *S. southlandiae* Larivière & Hoch) occur strictly on the South Island; they represent 50% of species of *Aka*, *Malpha* Myers, and *Semo*, and 67% of *Koroana*.

**Biology.** Very little is known about the habits of the majority of New Zealand species, except *Oliarus atkinsoni* which has been thoroughly studied because it is a known vector of plant disease. Most species appear to have a similar life cycle, with one egg stage, five nymphal stages, and a single generation each year. *Oliarus atkinsoni*, however, has a two-year life cycle. This is also the only species with described nymphs.

Female cixiids secrete a white woolly wax which is carried at the end of the abdomen and put around the eggs when they are laid in the soil. Nymphs develop around plant roots on which they feed. Adults are mostly active during the day. They are found on trees, shrubs, and amongst grasses, and generally stay on the most sun-exposed parts of their food plants where they can be collected by beating or sweeping. A number of species have been captured at night on low vegetation; this may indicate some degree of nocturnal activity.

In New Zealand, little is known about host plant requirements but there are definite habitat preferences. Most New Zealand cixiids inhabit forested or bush environments, including scrublands and shrublands, from coastal lowlands to the subalpine zone. Observed habitats and altitudinal ranges of genera are summarised in Table 2. The majority of genera are found in lowland to lower mountain mixed podocarp-broadleaf habitats. Information obtained so far on the genus *Aka* suggests a close association with *Nothofagus* forests, which may indicate an old line-

| Species | SD | NN | MB | KA | BR | WD | NC | MC | SC | MK | OL | CO | DN | FD | SL | SI | Offshore Islands |
|---|---|---|---|---|---|---|---|---|---|---|---|---|---|---|---|---|---|
| *Aka dunedinensis* | | | | | | | | | | | | | DN | | SL | | |
| *A. duniana* | SD | NN | | KA | | | | MC | | | | | | | | | |
| *A. finitima* | | | | | | | | | | | | | | | | | |
| *A. rhodei* | | | | | | | | | | | | | | | | | |
| *A. westlandica* | | NN | | | BR | WD | | MC | | | OL | CO | | FD | SL | SI | |
| *Chathamaka andrei* | | | | | | | | | | | | | | | | | CH |
| *Cixius inexspectatus* | | | | | | | | | | | | | | | | | |
| *C. punctimargo* | | | | | | | | | | | | | | | | | |
| *C. triregius* | | | | | | | | | | | | | | | | | TH |
| *Confuga persephone* | | NN | | | | | | | | | | | | | | | |
| *Huttia nigrifrons* | | | | | | | | | | | | | | | | | |
| *H. northlandica* | | | | | | | | | | | | | | | | | |
| *Koroana arthuria* | | | | | | | NC | MC | | MK | OL | | | FD | SL | SI | |
| *K. lanceloti* | | NN | MB | | BR | WD | | | | MK | OL | CO | | FD | | | |
| *K. rufifrons* | SD | NN | MB | KA | BR | | | MC | | | | | | | SL | | |
| *Malpha cockrofti* | | | | | BR | WD | | | | | | | | | | | |
| *M. muiri* | | | | | BR | | | | | | | | | | | | |
| *Oliarus atkinsoni* | | | | | BR | | | | | | | | | | | | |
| *O. oppositus* | SD | NN | MB | KA | BR | WD | NC | MC | SC | MK | OL | CO | DN | FD | SL | SI | |
| *Parasemo hutchesoni* | | | | | | | | | | | | | | | | | |
| *Semo clypeatus* | | NN | MB | | BR | | NC | MC | | | | | | | | | |
| *S. harrisi* | | NN | | | BR | WD | NC | | | MK | OL | | DN | FD | SL | SI | |
| *S. southlandiae* | | | | | | | | MC | | MK | | | DN | | SL | | |
| *S. transinsularis* | | NN | | | BR | | | | | | | | | | | | |
| *Tiriteana clarkei* | | | | | | | | | | | | | | | | | |

age. *Semo* is strictly a subalpine genus with largely allopatric species that are morphologically highly similar. This may indicate relatively recent speciation. The New Zealand alpine zone itself originated in the Pliocene; rapid and continuing speciation has been suggested for the evolution of the alpine biota (Kuschel 1975; Wardle 1991). *Koroana* is the only genus that spans a broad altitudinal range from lowlands to subalpine environments.

According to world literature, most cixiid species feed on a variety of plants, although some species have been reported to be oligophagous or monophagous. As for nymphs, the majority of records are from grasses (Wilson et al. 1994), but they have also been reported feeding on the roots of ferns (e.g., Zimmerman 1948), gymnosperms (e.g, Pinaceae, Sheppard et al. 1979), other monocotyledons (e.g., Agavaceae, Cumber 1952) and a number of dicotyledon families, including Asteraceae (Wilson et al. 1994). For adults, which feed above ground, most host records are from woody dicotyledons (Wilson et al. 1994), but some adult cixiids are reported from ferns (e.g., Zimmerman 1948), gymnosperms, and monocotyledons. Within monocotyledons, most records are from the Poaceae, Arecaceae, and Agavaceae. Most species have been recorded from a single host genus.

Table 3 results from the author's effort to compile a list of potential host plants from the literature, specimen labels, and her own fieldwork. In general, a method similar to Wilson et al. (1994) was used to minimise spurious collection or literature records. The following records were excluded: species collected by general sweeping; species observed but not feeding on the plant; species taken in general surveys in modified ecosystems; and those species whose taxonomic identity (in the literature) was questionable. Records of large numbers of newly emerged adults were deemed more indicative, and feeding records for nymphs were seen as the most reliable.

Similar tendencies to those observed elsewhere in the world can be seen in New Zealand. Most adults have been

**Table 2.** Altitudinal ranges and habitat types of New Zealand genera.

|  | Altitudinal ranges | Habitat types |
|---|---|---|
| *Aka* | Coastal lowlands - lower mountains. | Beech forests or mixed beech-podocarp-broadleaf forests shrublands & scrublands. |
| *Chathamaka* | Coastal lowlands. | Podocarp-broadleaf forests, shrublands & scrublands. |
| *Cixius* | Lowlands (inland & coastal). | As above. |
| *Confuga* | n/a | Caves. |
| *Huttia* | Lowlands (inland & coastal). | Podocarp-broadleaf forests. |
| *Malpha* | Lower mountains to subalpine environments. | Podocarp (incl. Kauri) - broadleaf, beech or mixed forests. |
| *Oliarus atkinsoni* | Lowlands (inland & coastal). | Flax marshes. |
| *O. oppositus* | Lowlands (inland & coastal) to subalpine environments. | Marshes, grasslands, grassy forest clearings, adventive pastures. |
| *Parasemo* | Lowlands to lower mountains. | Podocarp-broadleaf shrublands. |
| *Semo* | Mountains to subalpine environments. | Subalpine shrublands &. scrublands. |
| *Tiriteana* | Lowlands (inland & coastal). | Podocarp-broadleaf forests. |

found on woody dicotyledons, a lesser number in association with ferns, some on gymnosperms, and *Oliarus* on monocotyledons (mainly Poaceae and *Phormium* (Phormiaceae; previously in Agavaceae). The only information available for nymphs is for *Oliarus atkinsoni* which reproduces on *Phormium*. Information contained in Table 3 will hopefully guide future work on host plant preferences.

**Dispersal.** When local conditions become unsuitable, certain foreign species migrate long distances, usually at night. This phenomenon has never been reported in New Zealand where species seem to be able to leap or fly only short distances, presumably to escape danger or to move between plants. A tendency for brachyptery does not appear to be as marked as in other fulgoroid families, e.g.,

Delphacidae, and although genera such as *Aka* and *Chathamaka* gen. nov. have forewings that are somewhat shorter than other genera, they have hindwings that are usually fully developed. The author observed a number of individuals of *Chathamaka andrei* sp. nov. with their forewings welded together and slightly reduced hindwings.

**Economic importance.** Many species of Fulgoroidea are pests of cultivated plants around the world. Serious direct damage by Cixiidae, however, is rare in New Zealand. The greatest economic importance of Cixiidae is as vectors of phytoplasma plant diseases, e.g., *Oliarus atkinsoni,* on the New Zealand flax species.

**Table 3.** Most common plant associations.

| Genera | Associated plants |
| --- | --- |
| *Aka* | *Blechnum*[2], *Coprosma*[1]*, *Dracophyllum*[1], *Nothofagus*[1]* (nymphs in litter), *Pseudopanax*[1], *Schefflera*[1] |
| *Chathamaka* | *Blechnum*[2], *Coprosma*[1], *Dracophyllum*[1], *Melicytus*[1] |
| *Cixius* | Wide range of broadleaf shrubs[1] |
| *Confuga* | — |
| *Huttia* | *Agathis australis*[3], *Dacrydium cupressinum*[3], *Halocarpus kirkii*[3], *Podocarpus ferrugineus*[3], tree ferns[2] |
| *Koroana* | *Brachyglottis*[1], *Coprosma*[1], *Hebe** (mostly), *Melicytus*[1], *Olearia*[1] |
| *Malpha* | *Celmisia*[1], *Nothofagus*[1], *Olearia*[1], *Senecio*[1] |
| *Oliarus atkinsoni* | *Phormium*[4]** |
| *O. oppositus* | *Cyperaceae*[4], *Gramineae*[4], *Juncaceae*[4], *Poaceae*[4]* |
| *Parasemo* | ? |
| *Semo* | *Cassinia*[1], *Coprosma*[1], *Dracophyllum*[1], *Halocarpus*[3]*, *Hebe*[1]*, *Nothofagus*[1], *Olearia*[1] |
| *Tiriteana* | *Beilschmiedia*[1], *Coprosma*[1], *Carpodetus serratus*[1]* |

1 = Woody dicotyledons. 2 = Ferns. 3= Gymnosperms. 4 = Monocotyledons. * = Potential host. ** = Confirmed host.

## MORPHOLOGY AND TERMINOLOGY

The reader may acquire the elementary knowledge of adult cixiids morphology necessary to identify New Zealand taxa by reference to Figures 1–11. Other accounts of planthopper morphology can be found in O'Brien and Wilson (1985), the morphological terminology of which is generally adopted here in conjunction with that of recent taxonomic revisions (e.g., Van Stalle 1991). The term 'setiferous peduncles' is used to refer to the small tubercles set with setae, found along the forewing veins in the majority of species.

## METHODS AND CONVENTIONS

General working methods were the same as explained previously (Larivière 1995); they are not repeated here.

This study is based on the examination of over 4000 adult Cixiidae from over 700 New Zealand localities, and overseas reference material borrowed from the following institutions:

AMNZ  Auckland Institute and War Memorial Museum, Auckland.
ANIC  Australian National Insect Collection, Canberra, Australia.
ASCU  Agricultural Scientific Collections Unit, NSW Agriculture, Orange, Australia.
BMNH  The Natural History Museum, London, U.K.
BPNZ  B.H. Patrick collection, Dunedin (now deposited in Otago Museum, Dunedin).
CMNZ  Canterbury Museum, Christchurch.
FRNZ  Forest Research, Rotorua.
LUNZ  Lincoln University, Lincoln.
MONZ  Museum of New Zealand Te Papa Tongarewa, Wellington.
NZAC  New Zealand Arthropod Collection, Landcare Research, Auckland.
SAMA  South Australian Museum, Adelaide, Australia
UCNZ  University of Canterbury, Christchurch.

For locality records, area codes of Crosby *et al.* (1998) are listed from north to south and west to east. Each area is followed by collection localities listed alphabetically, with repository acronyms. A list of geographical coordinates for the main localities from which material was examined, is given in Appendix A.

Biological notes are based on an analysis of specimen label data. The native plants associated with New Zealand Cixiidae are listed in Appendix B, along with their family placement.

Very little specific or reliable information on distribution and biology could be found in the New Zealand literature (over 100 papers) which was thoroughly scanned.

The status, depositories, and full label data of all primary type specimens seen (and a summary of label information for secondary type specimens) are cited for each species. In the list of label data, different labels are separated by a solidus (/) and different lines on a label by a semicolon; all other punctuation is as it appears on the label.

**Collecting and preparation.** Adults of Cixiidae are generally collected by sweeping or beating shrubs and trees and, for some species, the low vegetation. Beating or handpicking from selected plants, especially host plants, yields invaluable biological information. Eggs and nymphs can be collected when the host plant is known by digging and sifting the soil around the roots of the host plant.

Adults are best preserved dry. Eggs and nymphs are usually kept in 70–75% ethanol. If a molecular study is intended, adults as well as immatures can be kept in 100% ethanol.

Preparation and curation of insects have been fully described by Walker and Crosby (1988). All specimens should be collected with the locality (including area code: Crosby et al. 1998, and geographical coordinates such as latitude and longitude), collection date, collector's name, and ecological data (e.g., general habitat and host plant).

Most features of the external morphology and the male genitalia can be viewed under an ordinary dissecting microscope. It is necessary to relax and dissect male specimens to study their genitalia.

Male genitalia can be dissected as follows. Pinned specimens are warmed for 5–10 minutes in soapy water or hot alcohol (70–75% ethanol). If the abdomen alone is to be used, this can be separated from the rest of the body by inserting a pin between it and the thorax, or, in difficult cases, by first relaxing the whole specimen in hot soapy water. Each specimen or abdomen is transferred to a watch glass half-filled with water (if soapy water was used) or ethanol, and the pygofer (genital capsule) is pulled away from the body using fine forceps and a micro-scalpel (needle tip from a 1.0 ml disposable hypodermic syringe). The pygofer is then warmed in very hot (almost boiling) ethanol for about 5 minutes, then transferred to another watch glass also containing ethanol. The anal tube, genital styles and aedeagus are detached and extracted from the pygofer in this solution, using fine forceps and a micro-scalpel. Dissected genitalia are subsequently stored in genitalia vials containing glycerine, and remounted on the pin below the relevant specimens.

Another method uses a 10% KOH solution to macerate the abdomen. This is done by putting the abdomen in a test tube containing a little KOH solution and by placing the test tube on a hot plate in a beaker containing water to warm up its content for 10–15 minutes (alternatively the abdomen can be left in cold KOH overnight). The rest of the dissection is done in water. This technique may be quicker for routine identification; KOH is a clearing agent that allows the examination of genitalia by transparency. If the purpose of the study is a taxonomic revision, it is preferable to extract the structures contained in the pygofer because they need to be viewed and illustrated in the exact same angle or position.

**Taxonomically relevant characters.** Detailed investigations of the external morphology, including the tymbal organ and the female abdomen, revealed very few characters that can be useful in species diagnosis. In most genera, species are morphologically similar, and intraspecific variation is high in most characters, e.g., forewing venation, shape and proportion of structures of the head and thorax.

The characters presented in the descriptions are subsets of the totality of adult characters studied, and represent the most important differences between, or variation amongst, closely related taxa. Characters or states of characters not mentioned in the species descriptions are as described in generic descriptions.

Descriptive measurements and counts were taken in the following manner: *vertex length* measured from tip of basal emargination to apex of vertex; *vertex width* taken at level of tip of basal emargination; *forewing length* taken from base to apex; *forewing width* measured at tip of clavus; *body length* measured from apex of head to tip of forewing, cited as a range with mean in parentheses.

Characters with the highest diagnostic value at the species level have been illustrated, including the most diagnostic aspects of the male genitalia. Most illustrations provided in this work represent the most commonly encountered state of a character. The user must allow some degree of variation when working with individual specimens, especially in the case of newly emerged adults that may still have a soft cuticle. It is not uncommon to see somewhat distorted genital structures in teneral individuals.

Keys are somewhat artificial. They are intended as an aid to identification, not a statement of the author's opinion on phylogenetic relations. Additional supporting characters (e.g., distribution) have sometimes been included between key couplets to help identification.

**Generic concept.** A genus should be a monophyletic group composed of one or more species separated from other genera by a decided gap. The phylogenetic framework to study Cixiidae, however, is insufficiently elaborated to test this hypothesis for New Zealand genera. Consequently, existing generic concepts have in general been accepted. In addition, two new genera are proposed for species not fitting the correlated character complex of species included in already described genera. Recognition of these generic taxa provides new hypotheses that will hopefully be tested by future students of the higher classification of Cixiidae; this must be done on a world basis or at least in an Australasian context.

A cladistic analysis, preferably integrating morphological and genetic information, is needed to determine the phylogenetic position of New Zealand genera within the Cixiidae. Only then can an attempt be made to decipher the evolutionary history of the New Zealand taxa, e.g., to confirm or reject the hypothesis that certain genera are Gondwana relicts, to reconstruct the sequence of speciation and colonization events, and to understand their evolution in general or that of their host plant relationships.

**Species concept.** The species concept used here is biological, inferred from morphological characters (especially male genitalia) hypothesised to constitute barriers to interbreeding and hence to gene flow between the different species (Larivière & Hoch 1998). This is corroborated, when possible, by geographic and biological information, but is not tested by genetic or ethological investigations. This species concept requires the assumption that reproductive (genetic) continuity or isolation among natural populations is evidenced by continuity or discontinuity in characters of external morphology and genital structures provided by the study of population samples.

As generally observed in Fulgoroidea, the most important characters to discriminate Cixiidae species are the male genital structures, particularly the aedeagus. In the majority of New Zealand genera, most external characters (e.g., forewing colour or venation, head or thorax morphology) are found to vary within species, or the range of their variation overlaps with that of closely related species, and for the most part similarities or differences in external morphology are not congruent with the study of genitalia. Accurate species identification is often virtually impossible without an examination of male genital structures. Therefore, in most cases, females can only be reliably identified by association with males. Fortuitously, identification is facilitated by the fact that New Zealand species are largely allopatric.

**Taxonomic arrangement.** Further study of Australasian Cixiidae is needed before phylogenetic relationships can be hypothesised, hence taxa are treated alphabetically in this monograph.

# KEY TO GENERA

**1** Eyes reduced, not visible in frontal view (Fig. 16). Vertex with expanded triangular anterior border (Fig. 104) ......................................... (p. 28) ... *Confuga* **Fennah**

—Eyes normally developed, visible in frontal view. Vertex without expanded triangular anterior border .... 2

**2**(1) Apical row of hind tarsomeres II (Fig. 4, ta II) with 6 teeth or more ......................................................... 3

—Apical row of hind tarsomeres II with less than 6 teeth ................................................................................. 4

**3**(2) Apical row of hind tarsomeres II with 13 teeth. Greyish or brownish species with head, thorax and abdomen deep, glossy black ......... (p. 44) ... *Oliarus* **Stål**

—Apical row of hind tarsomeres II with 6–9 teeth. Species differently coloured ......................................... 5

**4**(2) Hind tibiae with 3 lateral spines (Fig. 4, ls). Median carina of frons forked near midlength (Fig. 19, 20). Mesonotum with 5 longitudinal carinae (Fig. 108). Anal tube of male broad (Fig. 86) ............................ ........................................... (p. 35) ... *Malpha* **Myers**

—Hind tibiae without lateral spines. Median carina of frons simple, not forked near midlength (Fig. 23). Mesonotum with 3 longitudinal carinae (Fig. 3, lc; 111). Anal tube of male slender (Fig. 89) ................. ...........................................(p. 43) ... *Tiriteana* **Myers**

**5**(3) Forewings curved to fit closely around the body (Fig.101, 102).................................................... . 10

—Forewings not curved to fit closely around the body . .................................................................................. 6

**6**(5) Apical cells of each forewing with a well-defined dark spot ........................................................................ 7

—Apical cells of each forewing without a well-defined dark spot ................................................................. 8

**7**(6) Veins of forewings visibly covered with setiferous peduncles (as in Fig. 10) .. (p. 37) ... *Parasemo* **gen. nov.**

—Veins of forewings not visibly covered with setiferous peduncles (Fig. 11) ............... (p. 37) ... *Semo* **White**

**8**(6) Mesonotum with 5 longitudinal carinae (Fig. 105, 106). Median carina of frons, if present, forked near midlength. Anal tube of male as in Figure 84 ............................................. (p. 28) ... *Huttia* **Myers**

—Mesonotum with 3 longitudinal carinae (Fig. 3, lc; 103, 107). Median carina of frons simple, not forked near midlength (Fig.15, 18). Anal tube of male not as above ................................................................................... 9

**9**(8) Frons longitudinally bicoloured (brown medially, pale yellowish white laterally) (Fig. 18). Yellowish brown species ........................... (p. 30) ... *Koroana* **Myers**

—Frons not longitudinally bicoloured. Greenish species (fading to yellow when dead) ..................................... ........................................ (p. 24) ... *Cixius* **Latreille**

**10**(4) Hind tibiae with small lateral spines (as in Fig. 4, ls). Costa of each forewing with about 25 setiferous peduncles. Hind legs about as long as body ............. ............................. (p. 23) ... *Chathamaka* **gen. nov.**

—Hind tibiae without lateral spines. Costa of each forewing with 10–15 setiferous peduncles. Hind legs 1.5× longer than body ............ (p. 18) ... *Aka* **White**

## THE TAXA OF NEW ZEALAND CIXIIDAE

### Family CIXIIDAE

### Subfamily CIXIINAE, Tribe CIXIINI

#### Genus *Aka* White

*Aka* White, 1879: 216. Type species *Cixius finitimus* Walker, 1858: 81, by original designation.

**Description.** Brownish species, often with rather heavy blotches of dark brown approaching black on head, thorax, and forewings; characterised by long legs and forewings that are short and curved to fit the body.

Vertex approximately 0.9× as long as broad; transverse subapical keel narrowly M-shaped, connected to anterior margin by 2 short ridges; basal compartment with well-defined median keel; basal emargination V-shaped. Frons with median carina forked near midlength; median ocellus yellowish. Postclypeus with a median carina.

Pronotum with a median longitudinal carina; a pair of curved postocular carinae, one on either side of middle, their midportion reaching close to hind margin. Mesonotum with 3 longitudinal carinae (median one often evanescent). Forewings 2.5–2.7× longer than broad, usually hyaline, but often opaque and milky; veins with setiferous peduncles (more visible distally), costa with 10–15 such peduncles; apical cells 6–8 (usually 7) in number. Legs 1.5× longer than body; hind tibiae lacking lateral spines; hind tarsomere I with an apical row of 6–7 teeth, hind tarsomere II with 6 teeth.

**Male genitalia.** Externally visible portions uniform throughout the genus. Anal tube (Fig. 80) and left genital style (Fig. 56–60) as illustrated. Aedeagus (Fig. 33–37) with 3 spinose processes arising near the base of the flagellum (a right lateral process which is curved and 2 left lateral processes that are subrectilinear and variable in length, or 2 slender processes flanking a much shorter, hook-shaped one).

**Remarks.** This indigenous genus comprises seven species. Five species occur in New Zealand, three of which are newly described here. Both *Aka hardyi* Muir, 1931 and *A. tasmani* Muir, 1931, are only known from Tasmania.

The New Zealand species are almost identical externally. No consistent difference in external morphology or colour could be found. The shape of the left genital style is highly variable within species and does not constitute a useful diagnostic character. The only sound diagnostic features are those of the male aedeagus. Fortunately, identification is helped by the fact that species are largely allopatric.

The geographical distribution of *Aka* species follows very closely the distribution of *Nothofagus* species on mainland New Zealand (see Wardle 1991: 142, Fig. 7.25).

#### Key to species (males)

**1** Aedeagus (in ventral view, Fig. 37) with a slender uncurved spinose process on either side of a much shorter, hook-shaped median process. Left genital style with broad, squarish apex (Fig. 60) ......................... ................................... (p. 21) ... *Aka rhodei* **sp. nov.** *Geographical distribution (Map 4): Central North Island.*

—Aedeagus (in ventral view, Fig. 33–36) with a right lateral curved spinose process and 2 left lateral processes of variable shape and length. Left genital style with acutely rounded apex (Fig. 56–59) ....................... 2

**2**(1) Left lateral spinose processes of aedeagus subequal in length (Fig. 34, 36) .......................................... 3

—Left lateral spinose processes of aedeagus distinctly unequal in length (Fig. 33, 35) ............................. 4

3(2) Aedeagus (in ventral view, Fig. 34): Left lateral spinose processes short, not extending or slightly surpassing the curved portion of the right lateral process ........... ...................... . (p. 19) ... ***Aka dunedinensis* sp. nov.** *Geographical distribution (Map 4): Southeastern South Island.*

— Aedeagus (in ventral view, Fig. 36): Left lateral spinose processes longer, extending well beyond the curved portion of the right lateral process ........................... ............................. (p. 19) ... ***Aka duniana* (Myers)**

*Geographical distribution (Map 4):Across Cook Strait, from the northeastern South Island to the Wellington region (North Island), and on mountain ranges south of Lake Taupo.*

4(2) Aedeagus (in ventral view, Fig. 33): Right lateral spinose process semicircular, pointing towards the periandrium apex; left lateral processes thick ........... ............................. (p. 20) ... ***Aka finitima* (Walker)** *Geographical distribution (Map 4): Northern and central North Island, south to Wanganui.*

— Aedeagus (in ventral view, Fig. 35): Right lateral spinose process sinuate, pointing towards the periandrium base; left lateral processes thin ............. ........................... (p. 22) ... ***Aka westlandica* sp. nov**. *Geographical distribution (Map 4): South Island west coast, east to the Otago Lake area and western Central Otago, and Stewart Island.*

## *Aka dunedinensis* sp. nov.

Figures 34, 57; Map 4.

**Type data. Holotype**: Male (NZAC) labelled "NEW ZEALAND DN; Ross Ck. Reservoir; 13 Apr 1980 (handwritten); B. I. P. Barratt / HOLOTYPE; *Aka*; *dunedinensis* sp. nov.; Larivière 1999" (red label). Mounted on point; male genitalia dissected and stored underneath specimen in genitalia vial containing glycerol.

**Paratypes**: 4 females (2 MONZ, 2 NZAC) same data as holotype and bearing blue paratype labels.

**Description**. Adult brown, sometimes rather heavily blotched with dark brown, approaching black.

Vertex brown, usually slightly yellowish at middle; basal emargination shallowly V-shaped. Frons brown, mottled with yellowish along outer carinae, with a single pale spot on each side along frontoclypeal suture. Postclypeus dark brown.

Pronotum yellowish brown or brown, often mottled with darker brown. Mesonotum also brown. Forewings with a number of dark spots basally or base fumate, a paler brown cloud or a number of dark brown blotches on disc and sometimes also on distal third; outer veins alternately dark brown and whitish; stigma small, brown; costa with 4–6 brown marks; Sc+R forked basad of Cu; r-m located slightly basad of $M_{3+4}$; apical cells hyaline. Legs yellowish brown, usually with base and apex of femora infuscate and tibiae annulated with brown and yellowish brown (hind tibiae most often entirely yellowish brown); hind tarsomeres I and II, each with an apical row of 6 teeth.

Ventral sternites brown. **Male genitalia**. Anal tube as in type species, *A. finitima* (Fig. 80). Left genital style as in Fig. 57. Aedeagus (in ventral view, Fig. 34) with 3 spinose processes arising near the base of the flagellum (right lateral, sinuate process curving towards the left and extending beyond the left border of the periandrium; 2 short left lateral processes of subequal length, not or slightly surpassing the curved portion of the right lateral process).

Body length of males 4.92–5.29 (5.11) mm, of females 4.83–6.58 (5.70) mm.

**Geographical distribution** (Map 4). Southeastern South Island.

**Material examined.** A total of 18 specimens was seen from the following localities.
**South Island. DN**. Dunedin (AMNZ). Ross Creek Reservoir (NZAC). **SL**. Hokonui Hills, Dolamore Park (NZAC). Owaka (NZAC).

**Biology**. Lower montane species. Habitat poorly known. Two specimen labels bear the information *Nothofagus* forest (SL, Owaka) and ferns (SL, Dolamore Park). Most adults studied were collected in January, February, and April.

## *Aka duniana* (Myers)

Figures 36, 59; Map 4.

*Malpha duniana* Myers,1924: 323.
*Aka duniana* (Myers); –Fennah, 1975: 380.

**Type data. Holotype**: Male (NZAC) labelled "Type (circular red-bordered label) / Dun Mt; 3000'; R.J. Tillyard (handwritten) / abd. blackish (handwritten) /J.G. Myers Coll.; B.M. 1937-789. / (white card pygofer mounted between 2 round plastic pieces) / Holotype; Malpha; duniana; Myers (handwritten; first line at right angle along left border which is red)". Rather poor condition; double-mount; forewings spread out; abdomen missing.

**Allotype**: Female (NZAC) labelled "Dun Mt 2500ft; 17-4-21 (handwritten); A. Philpott / 164 (handwritten) / Allotype; Malpha; duniana; Myers (handwritten; first line at right angle along left border which is red)".

**Description**. Adult yellowish brown, sometimes rather heavily blotched with dark brown, approaching black.

Vertex pale yellowish brown, slightly darker brown at middle, with lateral borders and keels often distinctly paler yellow than remainder; basal emargination deeply V-shaped. Frons dark brown along median carina, fading to pale yellowish brown laterally, almost whitish near vertex. Postclypeus dark brown.

Pronotum pale yellowish brown, mottled with darker brown (whitish in some specimens). Mesonotum yellowish or pale brown at middle, darker brown laterally. Forewings with a number of dark spots basally or base fumate, a paler brown cloud or a number of dark brown blotches on disc and sometimes also on apical third; outer veins alternately dark brown and whitish; stigma small, brown; costa with 4–6 brown marks; Sc+R forked slightly basad of Cu; r-m located distad of $M_{3+4}$; apical cells hyaline. Legs whitish yellow or pale yellowish brown, usually with base and apex of femora infuscate and tibiae annulated with pale brown and yellowish or whitish (hind tibiae often almost completely unicoloured); hind tarsomeres I and II, each with an apical row of 6 teeth.

Ventral sternites pale brown. **Male genitalia**. Anal tube as in type species, *A. finitima* (Fig. 80). Left genital style as in Fig. 59. Aedeagus (in ventral view, Fig. 36) with 3 spinose processes arising near the base of the flagellum (right lateral sinuate process curving towards the left, extending beyond the left lateral border of the periandrium and pointing towards its base; 2 long left lateral processes of subequal length, extending well beyond the curved portion of the right lateral process).

Body length of males 4.50–5.33 (4.90) mm, of females 4.75–6.55 (5.42) mm.

**Geographical distribution** (Map 4). Across Cook Strait, from the northeastern South Island to the Wellington region (North Island), and on mountain ranges south of Lake Taupo.

**Material examined.** A total of 74 specimens was seen from the following localities.
**North Island. TO.** Ohakune (CMNZ). Tongariro National Park: Mount Ruapehu, Whakapapaiti (AMNZ). **HB.** Kaweka Forest Park: Kaweka Flats track (NZAC); Ngahere Catchment (NZAC). **WN.** Days Bay (NZAC). Tararua Forest Park/Range: Dundas Hut/Ridge; Mount Alpha; [Otaki] River Fork (NZAC).

**South Island. SD.** Mount Stokes, Okoha Saddle (LUNZ). Okiwi Bay (NZAC). Stephens Island (NZAC). Chetwode Islands, Te Kakaho Island (LUNZ). **NN.** Eves Valley, Palmers Bush (NZAC). Kaihoka Lakes (NZAC). Mount Burnett (LUNZ). Takaka (NZAC). Takaka Hill (NZAC). **KA.** Oaro (LUNZ). Puhi Puhi Valley (LUNZ). **MC.** Christchurch, Riccarton Bush (LUNZ).

**Biology**. The little information available suggests that *Aka duniana* occurs in and at the margins of coastal to montane (up to about 1000 m) *Nothofagus* or mixed forests, shrublands and scrublands, where it can be found by sweeping the ground cover or by beating shrubs. Adults collected from September to July, with December, January, and February appearing to be periods of peak abundance. Newly emerged adults collected in September and October.

Myers (1924) records for *Aka finitima* in the Tararua Range (now known as *A. duniana*): "Frequents lowland rain-forest, but is particularly numerous in subalpine scrub from 3,500 ft. to 4,000 ft." . . . "In July a freshly-emerged adult was found in leaf-mould, showing that the nymphal stadia are probably passed in a cryptozoic habitat".

**Remarks**. Myers (1924) originally described this species in the genus *Malpha*, with a comment to the effect that structures of the face and male genitalia were more reminiscent of *Aka*.

Myers apparently knew only of the type specimens from Dun Mountain (NN) and may not have suspected the presence of this species on the North Island. Examination of southern North Island material identified by previous workers as *Aka finitima*, suggests that the two species have been confused since the time of their description. See also the section Remarks under *Aka finitima*.

## *Aka finitima* (Walker)

Figures 10, 13, 33, 56, 80, 91,101; Map 4.
*Cixius finitimus* Walker, 1858: 81.
*Aka finitima* (Walker); –White, 1879: 216.

**Type data. Holotype**: Female (BMNH) labelled "Type (circle with green border) / CIXIUS FINITIMUS. (One line label folded in two) / N. Zeal.; 54.4". Good condition, double-mounted on card. [Note: The type locality of specimens collected by Lt-Colonel D. Bolton and bearing the British Museum accession number "54.4" is likely to be Auckland (see Dugdale 1988)].

**Description**. Adult (Fig. 101) brown, sometimes rather heavily blotched with dark brown, approaching black.

Vertex brown, usually slightly yellowish at middle; basal emargination shallowly V-shaped. Frons (Fig. 13) brown, mottled with yellowish along outer carinae, with a single pale spot on either side along frontoclypeal suture. Postclypeus dark brown.

Pronotum yellowish brown or brown, often mottled with darker brown. Mesonotum also brown. Forewings with a number of dark spots basally or base fumate, a paler brown cloud or a number of dark brown blotches on disc, sometimes also on apical third; outer veins alternately dark brown and whitish; stigma small, brown; costa with 4–6 brown marks; Sc+R forked basad of Cu; r-m located slightly basad of $M_{3+4}$; apical cells hyaline. Legs yellowish brown, usually with base and apex of femora infuscate and tibiae annulated with brown and yellowish; hind tarsomere I with an apical row of 6–7 teeth, hind tarsomere II with 6 teeth.

Ventral sternites brown. **Male genitalia**. Pygofer (Fig. 91), anal tube (Fig. 80), and left genital style (Fig. 56) as illustrated. Aedeagus (in ventral view, Fig. 33) with 3 spinose processes arising near the base of the flagellum (right lateral, semicircular process curving towards the left, extending beyond the left border of the periandrium, and pointing towards its apex; 2 rather thick, left lateral processes of unequal length, the shortest one of which is sinuate, the longest one, subrectilinear and extending to about 3/4 of the aedeagal length).

Body length of males 5.25–6.33 (5.52) mm, of females 5.62–6.80 (5.78) mm.

**Geographical distribution** (Map 4). Northern and central North Island, south to Wanganui.

**Material examined**. A total of 46 specimens was seen from the following localities.
**North Island. ND**. Poor Knights Islands, Tawhiti Rahi (NZAC). Tangihua Range, Horokaka (NZAC). **AK**. Huia, Destruction Gully (NZAC). Rangitoto Island (NZAC). **CL**. Kauaeranga Valley, Webb Creek (NZAC). Little Barrier Island, Summit track (NZAC). **BP**. Tarukenga (NZAC). Tikitapu/Blue Lake (NZAC). **TO**. Kaimanawa Forest Park, Clements Road, Te Iringa (NZAC). Ohakune (NZAC). **RI**. Mangahuia [Stream] (NZAC). Ruahine Forest Park: Colenso Trig (NZAC); Limestone Road end (NZAC). **WN**. Korokoro (NZAC). **WI**. Wanganui (NZAC).

**Biology**. Recorded from *Nothofagus* (*N. fusca*, *N. menziesii*) forests or mixed forests, shrublands, and scrublands, from coastal lowlands to about 900 m. Also collected on *Coprosma* (mostly) and *Xeronema*. Adult activity recorded from December to May.

**Remarks**. Myers (1924) redescribed this species based on 34 males and 29 females from "Wellington District, Tararua Range, Canterbury and Dunedin". Examination of material from these areas revealed, however, that the Wellington District, Tararua Range, and Canterbury specimens belong to *Aka duniana* and the Dunedin specimens to *Aka dunedinensis* sp. nov. Further study of material contained in New Zealand collections indicates that *Aka finitima* is a northern North Island species. On the other hand, *Aka duniana* is mostly a southern North Island–northern South Island species (see Map 4) with a disjunct distribution on the Central Volcanic Plateau of the North Island.

### *Aka rhodei* sp. nov.

Figures 37, 60; Map 4.

**Type data. Holotype**: Male (NZAC) labelled "NEW ZEALAND TO; Pureora; Waipapa Res, 570m; 26 Jan 1984; J. Hutchison / Malaise trap in; shrublands / HOLOTYPE; *Aka*; *rhodei* sp. nov.; Larivière, 1999 (red)". Mounted on point; male genitalia dissected and stored underneath specimen in genitalia vial containing glycerol.

**Paratypes** (7 males, 10 females) bearing blue paratype labels, with same data as holotype, except as follows: 2 males (MONZ, NZAC) and 3 females (NZAC), 17 Nov 1983; 3 males (NZAC), 24 Nov 1983; 1 female (MONZ), 15 Dec 1983; 1 female (NZAC), 29 Dec 1983; 3 females (MONZ), 5 Jan 1984; 1 male (MONZ), 23 Feb 1984; 1 male (MONZ), 1 Mar 1984; 1 female (NZAC), 8 Mar 1984.

**Description.** Adult yellowish brown, sometimes rather heavily blotched with dark brown. Vertex pale yellowish brown, slightly darker brown at middle, often with distinctly paler margins and carinae; basal emargination deeply V-shaped. Frons dark brown along median carina fading to pale yellowish brown laterally, almost whitish near vertex. Postclypeus dark brown.

Pronotum pale yellowish brown mottled with darker brown (whitish in some specimens). Mesonotum yellowish or pale brown at middle, darker brown laterally. Forewings with a number of dark spots basally or base fumate, a paler brown cloud or a number of dark brown blotches on disc, also sometimes on distal third; outer veins alternately dark brown and whitish; stigma small, brown; costa with 4–6 brown marks; Sc+R forked slightly basad of Cu; r-m located distad of $M_{3+4}$. Legs whitish yellow or pale yellowish brown, usually with base and apex of femora infuscate and tibiae annulated with pale brown and yellowish or whitish (hind tibiae often almost entirely unicoloured); hind tarsomeres I and II, each with an apical row of 6 teeth.

Ventral sternites pale brown. **Male genitalia**. Anal tube as in type species, *A. finitima* (Fig. 80). Left genital style as in Fig. 60. Aedeagus (in ventral view, Fig. 37) with 3 spinose processes arising near the base of the flagellum (2 slender uncurved processes, one on either side of a much shorter, hook-shaped median process).

Body length of males 4.92–6.25 (5.82) mm, of females 5.92–7.08 (6.70) mm.

**Geographical distribution** (Map 4). Limited to the central North Island.

**Material examined.** A total of 22 specimens was seen from the following localities.
**North Island. TO**. Pureora State Forest Park, Waipapa Reserve (NZAC). Waikato–Waipakihi Rivers junction (NZAC). **RI**. Mangahuia Stream (NZAC).

**Biology**. A lower montane species collected in the Pureora State Forest Park, in a shrubland terrace with pumice soil and a dominant vegetation of *Dracophyllum subulatum*, *Coprosma*, and *Pseudopanax*. Adults found from November to March, mostly in December and February. Newly emerged adults collected in November and December.

**Remarks**. *Aka rhodei* has a slightly different aedeagal configuration than other members of this genus. This species is named after my friend and colleague Birgit Rhode, who has contributed so much towards the completion of this work.

### Aka westlandica sp. nov.

Figures 35, 58; Map 4.

**Type data. Holotype**: Male (NZAC) labelled "NEW ZEALAND FD; Bauza I (handwritten); Thompson Sd (handwritten); Mar 1984 (handwritten); CF Butcher / ex Kamahi (handwritten) / HOLOTYPE; *Aka*; *westlandica* sp. nov.; Larivière, 1999 (red)". Mounted on point; male genitalia dissected and stored underneath specimen in genitalia vial containing glycerol.

**Paratypes** (7 males, 8 females) bearing blue paratype labels, with same data as holotype, except as follows: 1 male (NZAC), 3 females (MONZ), ex Schefflera; 2 males (MONZ), 1 female (NZAC), ex Mahoe; 1 male (NZAC), 1 female (NZAC), sweeping ex vegetation; 1 male (MONZ), ex Ascarina lucida; 1 male (NZAC), Weinmannia racemosa; 1 male (NZAC), ex Pseudopanax simplex; 1 female (MONZ), ex Pseudopanax crassifolium [sic]; 1 female (NZAC), ex Coprosma colensoi.

**Description.** Adult brown, sometimes rather heavily blotched with dark brown, approaching black.

Vertex brown, usually slightly yellowish at middle; basal emargination shallowly V-shaped. Frons brown, mottled with yellowish along outer carinae, with a single pale spot on either side along frontoclypeal suture. Postclypeus dark brown.

Pronotum yellowish brown or brown, often mottled with darker brown. Mesonotum also brown. Forewings with a number of dark spots basally or base fumate, a paler brown cloud or a number of dark brown blotches on disc and sometimes also on distal third; outer veins alternately dark brown and whitish; stigma small, brown; costa with 4–6 brown marks; Sc+R forked slightly basad of Cu; r-m located distad of $M_{3+4}$; apical cells hyaline. Legs annulated with brown and yellowish brown, except for hind tibiae which are most of the time entirely yellowish brown; hind tarsomeres I and II, each with an apical row of 6 teeth.

**Male genitalia**. Anal tube as in type species, *A. finitima* (Fig. 80). Left genital style as in Fig. 58. Aedeagus (in ventral view, Fig. 35) with 3 spinose processes arising near the base of the flagellum (right lateral sinuate process curving towards the left, extending beyond the left border of the periandrium and pointing towards its base; 2 rather thin left lateral processes of unequal length, the shortest one of which is sinuate, the longest one subrectilinear and extending to about 3/4 of the aedeagal length.)

Body length of males 5.00–5.83 (5.36) mm, of females 4.92–6.17 (5.82) mm.

**Geographical distribution** (Map 4). South Island west coast, east to the Otago Lake area and western Central Otago; Stewart Island.

**Material examined.** A total of 102 specimens was seen from the following localities.
**South Island. NN**. Collingwood (FRNZ). **BR**. Greymouth (NZAC), Boddytown (NZAC). Lake Moana (CMNZ). Moana (CMNZ). New River (NZAC). Porarari River (LUNZ). **WD**. Arawata River (LUNZ). Jackson Bay (NZAC). Lake Kaniere (LUNZ). Mount Aspiring National Park: Douglas Creek (NZAC); Roaring Billy Forest Walk (NZAC). Mount Hercules (LUNZ). Okuru (LUNZ). Poerua River Scenic Reserve (LUNZ). Westland National Park (NZAC): Canavans Knob (LUNZ); Lake Matheson (LUNZ). **MC**. Dean's Bush (CMNZ). **OL**. Hollyford Valley, Hollyford Road (NZAC). **CO**. Kawarau Gorge, Roaring Meg (NZAC). **FD**. Fiordland National Park: Bauza Island (NZAC); Doubtful Sound (LUNZ); Lake Manapouri (LUNZ, NZAC), Grebe Valley (LUNZ); Secretary Island (NZAC), Gut Bay (NZAC), track near Grono Bay (NZAC); Tutoko River (LUNZ). **SL**. Bluff Hill, Glory

Track (NZAC). Longwood Range, Orepuki Track (NZAC).
**Stewart Island.** Christmas Village (LUNZ). Codfish Island, Valley Track (NZAC). Freds Camp (LUNZ). Mason Bay (LUNZ). Mount Rakeahua (LUNZ) hut (LUNZ). Port Pegasus (NZAC). Port William (NZAC, LUNZ).

**Biology.** *Aka westlandica* occurs from coastal lowlands to lower montane areas, in and at the margins of *Nothofagus* or mixed forests, shrublands, and scrublands. Often collected on *Coprosma* in coastal areas or by sweeping ferns (e.g., *Blechnum*) in forested areas. Also recorded from the following native plants: *Ascarina lucida, Carpodetus serratus, Melicytus, Schefflera digitata, Pseudopanax crassifolius, P. simplex,* tree ferns, and *Weinmannia racemosa*. Males and females found on *Blechnum capense* at night (FD, March). Adults collected from October to May with December and February apparently being periods of peak abundance. Newly emerged adults recorded from October to December and in February.

**Remarks.** One teneral male individual collected from Central Otago (Kawarau Gorge; NZAC) has a slightly different aedeagus. It is tentatively treated as the same taxon until more material becomes available; aedeagal shape is often irregular in newly emerged adults.

## Genus *Chathamaka* gen. nov.

Type species *Chathamaka andrei* sp. nov., by present designation.

**Description.** Adults yellowish brown with darker brown head, pronotum, and mesonotum. Resembling *Aka* by body shape and forewings which are short and curved to fit the body.

Vertex approximately 0.9× as long as broad; transverse subapical keel V-shaped, sometimes narrowly M-shaped, connected to anterior margin by 2 short ridges; basal compartment with incomplete median keel; basal emargination deeply V-shaped. Frons with median carina forked near midlength; median ocellus yellowish. Postclypeus with median carina.

Pronotum with median longitudinal carina; a pair of curved postocular carinae, somewhat more elevated than in *Aka* and with their midportion not reaching as close to hind margin as in that genus. Mesonotum with 3 well-developed longitudinal carinae. Forewings 2.6–3.0× longer than broad, hyaline or slightly fumate with a few scattered brownish spots; veins with setiferous peduncles, costa with about 25 such peduncles; apical cells 7–8 in number, slenderer than in *Aka*. Legs 0.9–1.2× longer than body;

hind tibiae with 3, sometimes 4, very small, lateral spines; hind tarsomeres I and II, each with an apical row of 6 teeth.

**Male genitalia.** Anal tube as in Fig. 81. Left genital style L-shaped. Aedeagus (in ventral view) with 4, sometimes 3, spinose processes arising near the base of the flagellum.

**Remarks.** This Chatham Island taxon superficially resembles *Aka* but its morphology clearly does not fit the character complex found in New Zealand species already included in that genus. The high dissimilarity in external characters, especially legs and forewings, and in the male genitalia, especially the left genital style, aedeagus, and anal tube, suggests that this taxon does not share a common ancestor with *Aka* species.

The genus is currently monotypic, but four specimens from Chatham Island (Awatotara Forest tableland (NZAC); Waitangi (NZAC); Awatotara River mouth (NZAC)) may represent a different species. It will, however, be necessary to collect additional material before this can be established with certainty. The external morphology of these specimens as well as their anal tube, genital style, and pygofer fit the current description of *Chathamaka*, but the configuration of the male aedeagus is slightly different with only 3 spinose processes (2 on the left side and one on the right side, in ventral view).

## *Chathamaka andrei* sp. nov.

Figures 14, 38, 61, 81, 92, 102.

**Type data. Holotype**: Male (NZAC) labelled "NEW ZEALAND CI; Pitt I, Glory Scen Res; 22 .xi.1992; P. Syrett; beaten from Dracophyllum / HOLOTYPE; Chathamaka; andrei sp. nov.; Larivière, 1999 (red)". Mounted on point; male genitalia dissected and stored underneath specimen in genitalia vial containing glycerol.

**Paratypes** (18 males, 7 females) bearing blue paratype labels, with same data as holotype, except as follows: 1 female (LUNZ), "NEW ZEALAND CI; Pitt I, Glory Scen Res; 22.xi.1992; P. Syrett; beaten from vegetation"; 2 males (NZAC), 1 female (MONZ), "PITT ISLAND; Glory Scen Res; 18.I.1990; J.W. Early; C.A. Muir / swept in; regenerating; Dracophyllum; forest"; 7 males (3 MONZ, 4 NZAC), 2 females (NZAC), "PITT ISLAND; Canister Cove; Scientific Res; 12.I.1990; J.W. Early; broadleaf forest; remnant"; 5 males (3 LUNZ, 2 NZAC), 1 female, "PITT ISLAND; Canister Cove; Scientific Res; 12.I.1990; R.M. Emberson / beating; Myoporum"; 3 males, 1 female (LUNZ), "PITT ISLAND; Canister Cove; Scientific Res; 12.I.1990; R.M. Emberson / beating; vegetation"; 1 male, 1 female, "PITT ISLAND; Canister Cove; Scientific Res;

12.I.1990; C.A. Muir / remnant fern; broadleaf; forest".

**Description.** Adult (Fig. 102) yellowish brown with darker brown head, pronotum, and mesonotum.

Vertex brown with somewhat paler yellowish keels and margins; basal compartment also often pale yellowish on either side of middle. Frons (Fig. 14) brown, with paler yellowish brown carinae; median ocellus yellowish. Postclypeus slightly lighter brown than frons, not swollen, with pale yellowish brown median carina.

Pronotum brown with lighter yellowish brown carinae and margins. Mesonotum brown, somewhat paler yellowish brown apically. Forewings hyaline or slightly fumate, with a few brownish spots along costa, on disc, and on apex of clavus; veins yellowish to yellowish brown; stigma pale yellowish, sometimes whitish; Sc+R forked slightly basad of Cu; r-m located basad of $M_{3+4}$; apical cells usually hyaline or slightly fumate. Legs completely pale yellow with contrasting blackish claws and tibial spines.

Ventral sternites brown. **Male genitalia.** Pygofer (Fig. 92), anal tube (Fig. 81), and left genital style (Fig. 61) as illustrated. Aedeagus (in ventral view, Fig. 38) with 4 spinose processes arising near the base of the flagellum (a rather long process and a shorter one on either side of periandrium).

Body length of males 4.33–4.83 (4.51) mm, of females 5.00–5.58 (5.31) mm.

**Geographical distribution.** Endemic to the Chatham Islands.

**Material examined.** A total of 114 specimens was seen from the following localities.
   **Chatham Islands. Pitt Island** (CMNZ, NZAC). Canister Cove Scientific Reserve (LUNZ, MONZ, NZAC). Glory Bay (NZAC). Glory Scenic Reserve (LUNZ, MONZ, NZAC). Waipaua Scenic Reserve (LUNZ). Waipaua - Glory Bay (LUNZ). Kaingaroa (CMNZ).

**Biology.** Coastal lowlands. Collected on the vegetation in a regenerating *Dracophyllum* forest, shrubs (*Brachyglottis huntii*, *Coprosma chathamica*, *Melicytus chathamicus*, *Myoporum*), the edge of a remnant forest, and herbs and grasses hanging from coastal rocks. Adults collected from November to February; newly emerged adults, in November and January. Hindwings often reduced to about two-thirds the length of the forewings; probably incapable of flight.

**Remarks.** In a number of specimens the forewings were welded together and hindwings were reduced.

This species is named after my husband and colleague, André Larochelle, for his continued support in her career and for his infectious enthusiasm for life in general, and for nature in particular.

## Genus *Cixius* Latreille

*Cixius* Latreille, 1804: 310. Type species *Cicada nervosa* Linnaeus, 1758: 437, by subsequent designation of Curtis, 1837, British Entomology 4: pl. 673.

**Description** (New Zealand). Greenish species, fading to dull yellow when dead.

Vertex 0.6–0.8× as long as broad; transverse subapical keel subrectilinear, not connected to anterior margin by short ridges; basal compartment without median keel; basal emargination V-shaped. Frons with median carina simple (not forked) and sharp; median ocellus absent. Postclypeus with a sharp median carina.

Pronotum with a median longitudinal carina; a pair of curved postocular carinae, one on either side of middle, their midportion almost touching hind margin. Mesonotum with 3 longitudinal carinae (usually paler than remainder of mesonotum). Forewings about 3× longer than broad, usually hyaline; veins with setiferous peduncles, costa with 15–25 such peduncles; apical cells 9–12 in number. Hind tibiae with 3 lateral spines; hind tarsomere I with an apical row of 7–8 teeth, hind tarsomere II with 7–9 teeth.

**Male genitalia.** Externally visible portions variable within genus (New Zealand). Anal tube (Fig. 82, 83) and left genital style (Fig. 62–64) as illustrated. Aedeagus (Fig. 39–41) with 2–5 spinose processes variously placed along periandrium.

**Remarks.** This cosmopolitan genus has 3 species which are endemic to the New Zealand mainland. A fourth species, *Cixius kermadecensis* (not treated here) is endemic to the Kermadec Islands.

Additional information on the extensive synonymy associated with this genus is available in Metcalf (1936).

**Hybridization.** Two additional aedeagal configurations were observed in this genus: one in which the periandrium is flanked by a pair of forked spinose processes (Fig. 28), and another one in which the processes are shorter and simple (unforked) (Fig. 29).

These aedeagal configurations were observed 1) in localities where the species currently known as *C. aspilus* and *C. punctimargo* also occur, 2) in localities within their distribution ranges, or, 3) as individuals in a series clearly belonging to one or the other of these species. They are believed to represent occasional hybrid specimens between the two species.

Fifteen specimens of each type were examined from the following localities (see also Map 12).

*Aedeagus with forked spinose processes.* **ND.** Marlborough State Forest, Takitu Stream. Mount Camel Peninsula. Whangarei Heads. **AK.** Lynfield. Oratia. Waiheke Island. **CL.** Coromandel. Cuvier Island. Great Barrier Island. Little Barrier Island. Tapu-Coroglen road. Waikawau-Kennedy Bay. Waikawau Stream. Waiomu. **BP.** Tikitipu/Blue Lake. Waioeka Gorge. A number of specimens from Lynfield, Oratia, and Waikawau Stream also had one of the forked processes with a reduced arm.

*Aedeagus with simple spinose processes.* **ND.** North Cape. **CL.** Alderman Islands. Fantail Creek. Little Barrier Island. Mercury Islands. **BP.** Hicks Bay. Otanga Beach.

The male holotype of *C. aspilus* was dissected. Its aedeagal configuration corresponds that of the hybrid with two forked aedeagal processes. In this case Article 23 (h) of the current International Code of Zoological Nomenclature (1985) applies "A species-group name established for an animal later found to be a hybrid [Art 17 (1)] must not be used as the valid name of either of the parental species, even if it has priority over all other available names for them, but it may enter into homonymy."

Consequently, a new species (*Cixius inexspectatus* sp. nov.) is described for the organism previously identified as *Cixius aspilus* in the New Zealand literature.

Such hybrids may be the result of secondary intergradation between populations following shifts and fragmentation of species ranges, e.g., during the Pleistocene glacial cycles. The glacial period with its many changes in climate and sea level has had a major influence on the landforms of Auckland, Northland, and coastal areas of the Coromandels and the Bay of Plenty.

Obviously further collecting and some genetic work is required to prove the existence of these hybrids, but the author has little hesitation in making this assumption based on currently available data on morphology and distribution.

### Key to species

**1** Forewings almost clear; costa with 20–25 setiferous peduncles (as in Fig. 10). Left genital style of male as in Fig. 62. Aedeagus (in ventral view, Fig. 39) with 2 minute subapical teeth on right side, a long bifurcate subapical spinose process on left, near the base of the flagellum, and a pickaxe-shaped process on the right along the apical third of the periandrium ................
................ (p. 25) ... *Cixius inexspectatus* **sp. nov.**

—Forewings with scattered markings and several darker crossveins apically; costa with 15–25 setiferous peduncles. Left genital style of male and aedeagus different ................................................................ 2

**2**(1) Costa of each forewing with about 15 setiferous peduncles. Left genital style as in Fig. 63. Aedeagus (in ventral view, Fig. 40) with a single long spinose process arising subapically on right side, connected at its base to a curved plate which extends towards the left border of the periandrium and supports 2 shorter processes of subequal length. Average body length of males 4.92 mm, of females 5.38 mm ................................
................ (p. 26) ... *Cixius punctimargo* **Walker**

—Costa of each forewing with 20–25 setiferous peduncles. Left genital style as in Fig. 64. Aedeagus (in ventral view, Fig. 41) with 5 spinose processes (3 processes arising subapically near the base of the flagellum; 2 other processes arising near the base of the periandrium, one curving inwards towards the right, the other extending like a pointed rod towards the periandrium apex). Larger species, average body length of males 7.43 mm, of females 8.20 mm ...................
........................... (p. 27) ... *Cixius triregius* **sp. nov.**

*Distribution: Restricted to the Three Kings Islands.*

### *Cixius inexspectatus* sp. nov.

Figures 39, 62, 82, 93; Map 5.

*Cixius aspilus* in the sense of authors, not Walker, 1858 (invalid name based on hybrid; ICZN 23(h)).

**Type data. Holotype:** Male (NZAC) labelled "NEW ZEALAND TO; Pureora; Waipapa Res, 570m; 26 Jan 1984; J. Hutchison / Malaise trap in; Podocarps / HOLOTYPE; *Cixius*; *inexspectatus* sp. nov.; Larivière, 1999 (red)". Mounted on point; male genitalia dissected and stored underneath specimen in genitalia vial containing glycerol.

**Paratypes** (2 males, 5 females) bearing blue paratype labels, 1 female (NZAC) with same data as holotype, others as follows: 1 male (AMNZ), 5 Jan 1984, malaise traps in shrublands; 1 female (NZAC), 5 Jan 1984, malaise traps in podocarps; 1 male (MONZ), 12 Jan 1984, malaise traps in podocarps; 1 female (MONZ), 16 Feb 1984, malaise traps in podocarps; 2 females (1 NZAC, 1 AMNZ), 8 Mar 1984, malaise traps in podocarps.

**Description.** Adult pale greenish (fading to dull yellow in dried specimens), usually with greener forewing veins, carinae, and angles.

Vertex pale greenish, somewhat darker green laterally, about 0.8× as long as broad; surface of basal compartment deeply depressed (much more so than anterior compartment); basal emargination narrow, deeply V-shaped. Frons mostly pale greenish; median carina sharp; outer carinae slightly reflexed; frontoclypeal suture squarish or broadly arcuate. Postclypeus yellowish brown.

Pronotum pale greenish, narrow. Forewings hyaline, almost clear, sometimes slightly fumate along costa and apex of clavus; veins pale greenish, slightly darker and thicker towards apex; usually without darker cross veins apically; stigma yellowish green or whitish, narrow, elongate; costa with 20–25 setiferous peduncles; Sc+R forked basad of Cu; r-m at about same level as $M_{3+4}$; apical cells slender, 10–12 (usually 11) in number. Legs greenish yellow or yellowish brown; hind tibiae with 3 lateral spines tipped with brown; hind tarsomeres I and II, each with an apical row of 8 teeth.

Ventral sternites yellowish green to pale brownish yellow. **Male genitalia**. Pygofer (Fig. 93), anal tube (Fig. 82, and left genital style (Fig. 62) as illustrated. Aedeagus (in ventral view, Fig. 39) with 2 minute subapical teeth on right side, a long bifurcate subapical spinose process on left side near the base of the flagellum, and a pickaxe-shaped process on right side along apical third of the periandrium.

Body length of males 4.20–5.40 (4.92) mm, of females 4.90–6.10 (5.38) mm.

**Geographical distribution** (Map 5). Known from the northern half of the North Island.

**Material examined.** 143 specimens were seen from the following localities.
**North Island. ND.** Hen & Chicken Islands (AMNZ). Mangamuka Hill (AMNZ). Mount Camel Peninsula (AMNZ). Mount Orowhana (AMNZ). Tutukaka Harbour (NZAC). Whatupuke Island (NZAC). Poor Knights Islands: Aorangi I (NZAC) (Crater Bay (NZAC), Puweto Valley (NZAC)); Tawhiti Rahi (NZAC) (East Ridge (NZAC), North track (NZAC), Plateau (NZAC), Shag Bay (NZAC)). **AK.** Auckland (NZAC). Noises Islands, Motuhoropapa Island (NZAC). Waitakere Range: Piha, Beacon Point (NZAC). **CL.** Little Barrier Island (AMNZ, NZAC). Mercury Islands, Stanley I (NZAC). **TO** Pureora State Forest Park, Waipapa Reserve (NZAC).

**Biology.** Coastal lowlands and inland areas up to about 600 m in scrublands, shrublands, and open broadleaf or mixed broadleaf-podocarp forests. Collected frequently on *Melicytus* sp., *Coprosma* sp., and *Pseudopanax* sp. Adults found from September to March throughout the range of the species with peaks of abundance in December and January. Apparently univoltine.

## *Cixius punctimargo* Walker

Figures 15, 40, 63, 83, 103; Map 5.

*Cixius punctimargo* Walker, 1858: 81.
*Cixius interior* Walker, 1858: 82. **syn. nov.**

**Type data. Holotype**: Female (BMNH) labelled "Type (circular green-bordered label) / N. Zeal.; 54.4 / CIXIUS PUNCTIMARGO (One line label folded in two). Fair condition, mounted on point. [Note: The type locality of specimens collected by Lt.-Colonel D. Bolton and bearing the British Musuem accession number "54.4" is likely to be Auckland (see Dugdale 1988)].

**Description.** Adult (Fig. 103) pale olive green (fading to dull yellow in dried specimens), usually with slightly paler carinae, several dark crossveins apically on forewings, and brownish vertex and frons.

Vertex brownish, somewhat paler laterally, about 0.6× as long as broad; surface of basal compartment depressed (about as much as anterior compartment); basal emargination wider than in *C. inexspectatus*, less deeply V-shaped. Frons (Fig. 15) brownish, with paler carinae; outer carinae distinctly reflexed (more so than in *C. inexspectatus*); frontoclypeal suture squarish or slightly arcuate, outlined in blackish along base of postclypeus. Postclypeus also brownish, with somewhat paler median carina.

Pronotum often paler than remainder of body, narrow. Forewings hyaline, almost clear; veins greenish, slightly darker and thicker towards apex; several darker cross veins apically; stigma yellowish green or whitish, narrow, elongate; costa with about 15 setiferous peduncles; Sc+R forked basad of Cu; r-m at about same level as $M_{3+4}$; with 9–11 (usually 10) apical cells not as slender as in *C. inexspectatus*, 9–11 (usually 10) in number. Legs greenish yellow, or pale yellowish brown; hind tibiae with 3 lateral spines tipped with brown; hind tarsomere I with an apical row of 7 teeth, hind tarsomere II with 7–9 (usually 8) teeth.

Ventral sternites brownish, often nearly black. **Male genitalia**. Anal tube as in Fig. 83. Left genital style as in Fig. 63. Aedeagus (in ventral view, Fig. 40) with a single rather long spinose process arising subapically on right side of periandrium, connected at its base to a curved plate which extends towards the left border of the periandrium and supports 2 shorter processes of subequal length.

Body length of males 4.20–5.40 (4.92) mm, of females 4.90–6.10 (5.38) mm.

**Geographical distribution** (Map 5). Known from the northern half of the North Island.

**Material examined.** 111 non-type specimens were seen from the following localities.
**North Island. ND.** Kaitaia (NZAC). Mangamuka Hill (AMNZ). Marlborough State Forest, Takitu Stream (AMNZ, NZAC). Opononi (AMNZ), Waima State Forest (AMNZ). Waipoua State Forest (LUNZ, NZAC). **AK.** Rangitoto Island (NZAC). Titirangi (NZAC). Waitakere Range: Fairy Falls (NZAC); Piha (NZAC); Sharps Bush (NZAC); Walker Bush track (NZAC). **CL.** Alderman Islands, Ruamahuanui Island (NZAC). Cuvier Island (AMNZ). Kirikiri Saddle (NZAC). Little Barrier Island: Summit track (AMNZ); Thumb track (NZAC). Mercury Islands, Red Island (NZAC). Tapu Hill (FRNZ, NZAC). Waitete Bay (NZAC). **BP.** Hicks Bay (NZAC). Mount Tarawera (NZAC). Papatea (NZAC). Rereauira (NZAC). Waioeka Gorge (AMNZ). **GB.** Urewera National Park: Lake Waikaremoana, Whaitiri Point (NZAC).

**Biology.** Altitudinal range, habitat, and associated plants as in *C. inexspectatus*. Adults collected from September to January, with peak abundance in November. Apparently univoltine.

**Remarks.** The female holotype of *Cixius interior* (BMNH) was seen. Its external morphology suggests that it is conspecific with *C. punctimargo*.

## *Cixius triregius* sp.nov.

Figures 41, 64; Map 5.

**Type data. Holotype**: Male (NZAC) labelled "Castaway; Camp / Three Kings Is; Great I. Nov. 70; N.Z. Ent. Div. Exp. / G. Kuschel / *Cixius*; *triregius*; sp. nov.; Larivière, 1999 (red label)". Good condition although forewings are somewhat shrivelled because specimen is teneral; male genitalia dissected, stored underneath specimen in genitalia vial containing glycerol.
**Paratypes** (2 males, 3 females): 2 females (1 NZAC, 1 AMNZ) labelled as holotype; 1 male (AMNZ), 1 female (MONZ) labelled "Castaway; Camp / Three Kings Is; Great I. Nov. 70; N.Z. Ent. Div. Exp. / J. McBurney / Plectomyrta (handwritten)"; 1 male (NZAC) labelled "THREE KINGS IS; Castaway Camp; 28 Nov 1970; G.W. Ramsay (handwritten)".

**Description.** Adult greenish with brown (fading to dull yellowish brown in dried specimens), usually with slightly darker carinae, several dark crossveins and a number of spots near middle and apex on forewings, especially in female.

Vertex somewhat paler medially, about as long as broad; lateral margins more elevated along basal compartment than along remainder of vertex; surface of basal compartment deeply depressed (much more so than anterior compartment), without a median keel; basal emargination shallowly V-shaped. Frons greenish or brownish, with paler, distinctly reflexed outer carinae which are subrectilinear or slightly concave at midlength; frontoclypeal suture arcuate, thinly outlined in brownish along base of postclypeus. Postclypeus not swollen, brownish, with well-developed, somewhat paler, sharp median carina.

Pronotum often paler than remainder of body, narrow, with a median longitudinal carina; a pair of curved postocular carinae present on either side of middle, their midportion almost touching hind pronotal margin. Mesonotum green or brownish, somewhat darker green laterally, with 3 longitudinal carinae (usually paler than remainder of mesonotum). Forewings 2.5–3× longer than broad, hyaline; veins greenish, slightly darker and thicker apically; several dark cross veins and a number of dark spots near middle and towards apex; stigma greenish or brownish with whitish edge, narrow, elongate; costa with 20–25 peduncles; Sc+R forked basad of Cu; r-m at about same level as $M_{3+4}$; apical cell 10–12 (usually 11) in number. Legs greenish yellow, or pale yellowish brown; hind tibiae with 3 lateral spines tipped with brown; hind tarsomere I with an apical row of 7 teeth, hind tarsomere II with 8 teeth.

Ventral sternites brownish. **Male genitalia.** Left genital style as in Fig. 64. Aedeagus (in ventral view, Fig. 41) with 5 spinose processes: 3 spinose processes arising subapically near the base of the flagellum and 2 processes arising near the base of the periandrium (one curving inwards toward the right side, the other extending like a pointed rod towards the periandrium apex).

Body length of males 6.90–8.60 (7.43) mm, of females 7.90– 8.77 (8.20) mm.

**Geographical distribution** (Map 5). Restricted to the Three Kings Islands.

**Material examined.** A total of 11 specimens was seen from the following localities.
**Three Kings Islands.** Great Island (NZAC): Castaway Camp (AMNZ, MONZ, NZAC), Summit Ridge (NZAC); Tasman Valley (NZAC). West Island (NZAC).

**Biology.** Coastal scrublands and shrublands. Associated plant records: *Myoporum laetum* and *Solanum aviculare* var. *albiflorum*. Adults and tenerals collected in November and January.

## Genus *Confuga* Fennah, 1975

*Confuga* Fennah, 1975: 377. Type species *Confuga persephone* Fennah, 1975: 379, by original designation.

This monotypic, cave-dwelling genus is only known from the type series and some additional nymphs from the Northwest Nelson (NN) area. Fennah (1975) offered an excellent description of the type species, the main characteristics of which are summarised below.

### *Confuga persephone* Fennah

Figures 16, 30–32, 104.

*Confuga persephone* Fennah, 1975: 379.

**Type data. Holotype**: Male (NZAC): Council Cave, Takaka, Nelson Prov., New Zealand, 4.iii.1974, L. McRae & J. McBurney.
**Allotype**: Female (NZAC), from same locality as holotype, 12.xii.1973, J. McB.

**Description.** Adult lightly pigmented, pale yellowish brown with membranous areas creamy white. Reduced, unpigmented eyes and ocelli. Vertex 1.5× longer than broad; basal compartment without median keel; apical compartment with expanded, triangular anterior border; basal emargination rectilinear. Frons (Fig. 16) with median carina entire (not forked). Pronotum quite long compared with other New Zealand genera, without a median longitudinal carina; a pair of curved postocular carinae, one on either side of middle, which are short and not reaching lateral margins. Mesonotum with 3 longitudinal carinae. Forewings broadly oval, somewhat reduced, with thick, brown veins with setiferous peduncles, costa with about 10 such peduncles; apical cells 9 in number. Hind tibiae with 3–4 minute, lateral spines; hind tarsomeres I and II, each with an apical row of 5 teeth.
**Male genitalia** (Fig. 30–32), after Fennah (1975). Pygofer, anal tube, and genital style as illustrated. Aedeagus (in dorsal view) with a short, stout, strongly sinuate spinose process on the right near the apex, a long oblique rib on the left side, descending from the apex of the periandrium towards its base and emerging as a spinose process directed cephalad; flagellum deeply bifid with each arm tapering into a spinose process, the lower longer than the other and curving to left.

**Geographical distribution**. Known only from the type locality, Council Cave (NN).

**Material examined**. Types only.

**Biology**. Poorly known. Adults collected in a limestone cave in December and March, and nymphs in October, March, April and June. Food source (Millar 1998): roots of trees which penetrate well beneath the limestone surface. Egg described by Fennah (1975).

**Remarks.** Nymphs of a very similar planthopper have been found in a cave at Paynes Ford Scenic Reserve, several kilometres to the west of Council Cave. They appear to be of the same genus but an adult male has yet to be found to establish if this is the same species. What does seem likely is that the same or similar species will eventually turn up in other limestone sites in the same region (Millar 1998).

## Genus *Huttia* Myers

*Huttia* Myers, 1924: 321. Type species *Huttia nigrifrons* Myers, 1924: 321, by original designation.

**Description.** Yellowish brown species, sometimes with an olive tinge and a few pale to dark brown marks on head, thorax and forewings, especially on apical crossveins, clavus, and along costa.
Vertex approximately 0.6× as long as broad; transverse subapical keel broadly arcuate, usually connected to anterior margin by 2 ill-defined, short ridges; basal compartment with or without median keel; basal emargination broadly arcuate to shallowly V-shaped. Frons without median carina or with a more or less developed one forked near its midlength; median ocellus concolorous with surroundings. Postclypeus lacking a median carina.
Pronotum with a median longitudinal carina; a pair of curved postocular carinae, one on either side of middle, their midportion reaching close to hind margin. Mesonotum with 5 longitudinal carinae (inner pair sometimes ill-defined). Forewings about 3× longer than broad, usually hyaline, often with 3–4 dark marks along costa; veins with setiferous peduncles, costa with 15–25 such peduncles; apical cells 8–9 in number. Hind tibiae with 3–5 lateral spines; hind tarsomeres I and II, each with an apical row of 7 teeth.
**Male genitalia**. Externally visible portion uniform throughout the genus. Anal tube (Fig. 84) and left genital style (Fig. 65, 66) as illustrated. Aedeagus (Fig. 42, 43) with 3 medium-length spinose processes (c. aedeagal length) arising subapically near the base of the flagellum, the median process shorter, sinuate or hook-shaped.

**Remarks.** An endemic genus with two species. Examination of the female holotype of *H. harrisi* revealed it to be conspecific with *Semo westlandiae*, a species recently described by Larivière & Hoch (1998). Therefore, *H. harrisi* is transferred to the genus *Semo* Myers as a senior synonym of *S. westlandiae* (see page 40).

### Key to species

1 Frons almost entirely shining dark brown or blackish, with a transverse band of white on frontoclypeal suture, followed by a transverse nearly semicircular band of shining dark brown or blackish on postclypeus (Fig. 17); lacking a median carina. Hind tibiae with 3 robust lateral spines. Left genital style of male as in Fig. 65. Short median spinose process of aedeagus sinuate and joined to left lateral process (Fig. 42) ...............
................ (p. 29) ... ***Huttia nigrifrons* Myers**

—Frons entirely pale yellowish with outer carinae outlined in brown; median carina more or less developed, forked near midlength. Hind tibiae with 4–5 small lateral spines. Left genital style of male more broadly shaped as in Fig. 66. Short median spinose process of aedeagus hook-shaped, not joined to left lateral process (Fig. 43) (p. 30) ... ***Huttia northlandica* sp. nov.**

## *Huttia nigrifrons* Myers

Figures 17, 42, 65, 84, 94, 105; Map 6.
*Huttia nigrifrons* Myers, 1924: 321.

**Type data. Holotype**: Female (NZAC) labelled "Type (circular red-bordered label) / Upper Hutt.; 5.12.20 (handwritten) / 112a (handwritten) / J.G. Myers Coll.; B.M. 1937-789. / Holotype; Huttia; nigrifrons; Myers (handwritten; first line at right angle along red left margin)". Poor condition; abdomen and right forewing missing; tip of left forewing damaged; right mid and hind legs only with femur.

**Description.** Adult (Fig. 105) rather large and robust, generally dark yellowish brown to chocolate brown, sometimes tinged with olive, with sharply defined dark brown or blackish marks on head and forewings.

Vertex pale yellowish brown with dark brown or blackish at middle; basal compartment much more depressed than anterior compartment, without median keel or, more rarely, with trace of an incomplete one; basal emargination broadly arcuate (shallowly U-shaped). Frons (Fig. 17) almost entirely shining dark brown or blackish, a transverse band of white on frontoclypeal suture, followed by a transverse nearly semicircular band of shining dark brown or blackish on postclypeus, remainder of postclypeus pale yellowish brown; without a median carina.

Pronotum yellowish with dark brown or blackish on disc. Mesonotum yellowish brown, somewhat darker brown laterally, usually with paler longitudinal carinae. Forewings hyaline, almost clear, with a few dark marks along costa and larger blotches on clavus and on apical crossveins, also often fumate or opaque brown at base and across disc; veins yellowish brown, almost transparent or alternately marked with yellowish and brown or blackish; costa with about 25 setiferous peduncles; stigma yellowish brown; Sc+R forked basad of Cu; r-m located at same level as $M_{3+4}$. Legs yellowish brown, sometimes with apices of femora paler; hind tibiae with 3 robust lateral spines (apical one usually more developed).

**Male genitalia.** Pygofer (Fig. 94), anal tube (Fig. 84), and left genital style (Fig. 65) as illustrated. Aedeagus (in ventral view, Fig. 42) with short median spinose process sinuate and joined to left lateral process.

Body length of males 4.30–5.30 (4.9) mm, of females 4.71–5.80 (5.2) mm.

**Geographical distribution** (Map 6). Throughout the North Island.

**Material examined.** A total of 20 specimens was seen from the following localities:
**North Island. ND.** Motuti River (AMNZ). North Cape, Ngaroku Stream (NZAC). Te Paki Coastal Reserve (NZAC). Te Paki Trig track (NZAC). Tutukaka Harbour (NZAC). Waikare River (AMNZ). Waipoua State Forest, Te Matua Ngahere (NZAC). Whangarei Heads, Mount Manaia (NZAC). **AK.** Huia, Destruction Gully (NZAC). Hunua Range, Mangatangi Valley (NZAC). Waitakere Range: Sharps Bush (NZAC). **CL.** Little Barrier Island: Bunkhouse area (NZAC); Thumb track (NZAC). **BP.** Kaimai-Mamaku Forest Park: Wright Road end, Aongatete Lodge track (NZAC). Lottin Point Road, Waenga Bush (NZAC). Orete Forest, Te Puia Hut (NZAC). Urewera National Park: Waimana Valley (NZAC). Waiaroho (NZAC). **TO.** Ohakune (UCNZ). **WN.** Upper Hutt (NZAC).

**Biology.** This species inhabits lowland mixed podocarp-broadleaf forests and their margins, where it is mostly found on podocarp trees. Also beaten from costal shrubs, swept from podocarps (*Dacrydium cupressinum, Halocarpus kirkii*), collected on *Prumnopitys ferruginea* (branch trap), and once found on young tree ferns. Adults collected from October to May, but mostly in October and November. Teneral individuals found in mid October and in April-May (kauri forest).

**Remarks.** Although fairly well distributed on the North Island, this endemic species is never abundant locally. Apart from *Semo*, it is the only other New Zealand Cixiidae for which a close association with Podocarpaceae is suspected.

## *Huttia northlandica* sp. nov.

Figures 43, 66, 106; Map 6.

**Type data. Holotype**: Male (NZAC) labelled "NEW ZEALAND ND; Omahutu [Omahuta] SF; 15 Jul 1974; J.S. Dugdale (handwritten) / sweeping; at night (handwritten) / HOLOTYPE; *Huttia*; *northlandica* sp. nov.; Larivière, 1999 (red label)". Male genitalia dissected, stored underneath specimen in genitalia vial containing glycerol.
**Paratypes** (4 males, 4 females) bearing blue paratype labels, with same data as holotype except as indicated below: 1 male (MONZ), 1 female (NZAC), with same data as holotype; 2 females (AMNZ, NZAC), "NEW ZEALAND ND; Warawara SF; 10 Oct 1974; J.C. Watt; beaten"; 1 male (MONZ), "Waipoua; S.F.; Oct. 67; Auckland/ J.C. Watt"; 1 male (AMNZ), "NEW ZEALAND ND; SH 12; Waipoua SF; 20 Sep 1977; J.K. Barnes"; 1 male, 1 female (on same pin; NZAC), WAIPOUA F.; 3.10.57; R.A. Cumber".

**Description.** Adult (Fig. 106) dull yellowish brown, with a few evanescent pale brown marks on forewings.
Vertex yellowish brown; basal emargination shallowly V-shaped. Frons entirely pale yellowish with outer carinae thinly outlined in brown; median carina more or less developed, forked near midlength. Postclypeus pale yellowish brown, not swollen, lacking a median carina.
Pronotum yellowish. Mesonotum yellowish brown, often paler at sides. Forewings almost clear; veins yellowish brown throughout or alternated yellowish and brown; costa with 15–20 setiferous peduncles; stigma brownish; Sc+R forked much basad of Cu; r-m much basad of $M_{3+4}$. Legs yellowish or fumate; hind tibiae with 4–5 small acute lateral spines (sometimes with a number of minute intercalary spines).
**Male genitalia.** Anal tube as in type species, *H. nigrifrons*. Left genital style as in Figure 66. Aedeagus (in ventral view, Fig. 43) with short median process hook-shaped, not joined to the left lateral process.
Body length of males 4.10–4.40 (4.2) mm, of females 4.60–5.00 (4.76) mm.

**Geographical distribution** (Map 6). Western Northland.

**Material examined.** Type specimens and one specimen from ND, Matarua Forest, Waioku Coach Road track (NZAC).

**Biology.** Adults collected in July (swept at night), September, and October. Habitat unknown.

## Genus *Koroana* Myers

*Koroana* Myers, 1924: 319. Type species *Cixius rufifrons* Walker, 1858: 83, by present designation and resurrection from synonymy with *Cixius interior* Walker, 1858.

**Description.** Distinctive genus, comprising yellowish brown slender species, often tinged with mossy green or reddish, and with frons longitudinally bicoloured.
Vertex approximately 0.6× as long as broad; transverse subapical keel subrectilinear or slightly arcuate, not connected to anterior margin by small ridges; basal compartment with median keel more or less defined; basal emargination V-shaped. Frons (Fig. 18) brownish medially, ivory to pale yellow laterally, often tinged with green; median carina simple (not forked); median ocellus apparently absent. Postclypeus yellowish brown to dark brown, not swollen, with a median carina (sometimes evanescent).
Pronotum with a median longitudinal carina (often weakly defined); a pair of curved postocular carinae, one on either side of middle, subparallel to hind margin. Mesonotum with 3 longitudinal carinae. Forewings 2.5–3.5× longer than broad, hyaline, with or without an irregular pattern of pale or dark brown spots coalescing into a poorly defined transverse band between costa and distal third of clavus; veins with setiferous peduncles, costa with 14–20 such peduncles; apical cells 9–11 in number. Hind tibiae with 3 immovable lateral spines; hind tarsomere I with an apical row of 6 teeth, hind tarsomere II with 8 teeth.
**Male genitalia.** Externally visible portions uniform throughout the genus. Anal tube (Fig. 85) and left genital style (Fig. 68–70) as illustrated. Aedeagus (Fig. 44–46) with 2 subapical spinose processes arising near the base of the flagellum (dorsolateral process short and subrectilinear or slightly curved; ventral process longer, curved dorsad, forked and ending in 2 sickle-shaped spinose processes near the midportion of the periandrium).

**Remarks.** The author recently published a taxonomic review of this genus (Larivière 1997b). The synonymy, type data, and slightly modified versions of the key to species and descriptions are included here in order to make this faunal review as comprehensive as possible. Additional information is provided on material examined, geographical distribution, and biology.

The female holotype of *Cixius interior* Walker, 1858 (p. 82) was obtained from the Natural History Museum (London). It is clearly not conspecific with *Koroana helena* Myers, 1924 and *Cixius rufifrons* Walker, 1858 as previously stated by Myers (1927), nor does it belong to *Koroana*.

On the other hand, examination of the holotype of *C. rufifrons*, including the dissection of the male genitalia which apparently was not seen by Myers (1927), confirmed its synonymy with *K. helena*. Consequently, *C. rufifrons* is here considered to be the real type species of *Koroana*.

**Hybridization**. Hybrid individuals (*K. lanceloti* x *K. rufifrons*; *K. lanceloti* and *K. arthuria*) can occasionally be found in the zone of geographic overlap between species (See Larivière 1997: 222).

### Key to species (based primarily on males)

1 Aedeagus (in dorsal view, Fig. 45) with a long (approximately half of aedeagal length), subrectilinear, dorsolateral spinose process arising subapically near the base of the flagellum and 2 slender, short, sickle-shaped, spinose processes near the midlength of the periandrium, the dorsally directed process slightly longer than the more narrowly coiled ventral process. Forewings usually (90% of specimens) clear, rarely with an ill-defined transverse band of pale brown spots between costa and clavus; apical cells hyaline ..........
.................... (p. 33) ... ***Koroana rufifrons*** **(Walker)**
*Distribution (Map 7): throughout the North Island; northern and southeastern South Island.*

—Aedeagus (in dorsal view, Fig 44, 46) with a shorter (less than half of aedeagal length), curved, dorsolateral spinose process arising subapically near the base of the flagellum and 2 robust or differently orientated, sickle-shaped, spinose processes. Forewings with a well-defined transverse band of dark spots between costa and clavus; at least one apical cell opaque dark brown .................................................................... 2

2(1) Aedeagus (in dorsal view, Fig. 44) with a very short (less than one-quarter of aedeagal length), sinuate, dorsolateral spinose process arising subapically near the base of the flagellum and 2 rather thick, short, sickle-shaped, spinose processes visible near the midlength of the periandrium, the dorsally directed process slightly longer than the ventral process ....................
...................... (p. 31) ... ***Koroana arthuria*** **Myers**
*Distribution (Map 7): Arthurs Pass (NC), Cass (MC), and southern South Island; Stewart Island.*

—Aedeagus (in dorsal view, Fig. 46) with one moderately long (approximately one-third of aedeagal length), sinuate, dorsolateral, spinose process arising more apically near the base of the flagellum and 2 thin, short, sickle-shaped, spinose processes near the midlength of the periandrium, the dorsally directed process much longer than the less narrowly coiled ventral process
.................... (p. 32) ... ***Koroana lanceloti*** **Larivière**
*Distribution (Map 7): mostly the South Island west coast and east of the Southern Alps around Otago Lakes and Mount Cook (MK).*

### *Koroana arthuria* Myers

Figures 44, 68, 85, 95; Map 7.

*Koroana arthuria* Myers, 1924: 320.

**Type data. Holotype**: Male (NZAC, indefinite loan from BMNH) labelled "Type (circular red-bordered label) / Arthur's Pass; 12.XI.22; J.G. Myers; 2500' (handwritten) / Emerged; 24.XI.22 (handwritten) / J.G. Myers Coll.; B.M. 1937-789. / Holotype; Koroana; arthuria; ♂ Myers (handwritten; first line at right angle along left border, which is red)." Very good condition; mounted on card point.

The allotype, apparently labelled as the holotype, could not be located.

**Description**. Adult brown, often with a reddish tinge; forewings with dark brown spots arranged in an irregular transverse band between costa and distal third of clavus.

Vertex brown, with basal compartment often paler; basal emargination deeply V-shaped (more deeply incised than in *K. rufifrons*).

Pronotum pale yellowish brown to brown, often darker laterally. Mesonotum yellowish white, pale brown, or, more rarely, mossy green medially, darker brown laterally, often tinged with reddish. Forewings hyaline, sometimes slightly infumate or milky; veins yellowish brown, often nearly black; stigma brown; costa with 17–20 peduncles; Sc+R usually forked distad of Cu, more rarely at same level as Cu; r-m usually located slightly distad of $M_{3+4}$ or at same level; A1 and Y-vein often whitish with brown spots near distal third of clavus; apical cells usually 9 (sometimes 10) in number, with 1 or 2 partly or entirely opaque dark brown; tegula slightly darker than pronotum. Legs brown to almost black, with hind tibiae sometimes yellowish brown; fore and mid tibiae annulated blackish and ivory or yellowish; hind tibiae with 3 immovable lateral spines.

Ventral sternites brown to blackish. **Male genitalia.** Pygofer (Fig. 95), anal tube (Fig. 85), and left genital style

(Fig. 68) as illustrated. Aedeagus (in dorsal view, Fig. 44) with one very short (less than one-quarter of aedeagal length), sinuate, dorsolateral, spinose process located subapically near the base of the flagellum and 2 thick, short, sickle-shaped, spinose processes visible near the midlength of the periandrium, the dorsally directed process slightly longer than the ventral process.

Body length of males 4.45–5.36 (4.94) mm, of females 4.73–5.33 (4.09) mm.

**Geographical distribution** (Map 7). Southernmost areas of the South Island, Stewart Island, and one population from Arthurs Pass (NC) and Cass (MC).

**Material examined.** A total of 196 specimens was seen from the following localities. **South Island**. **MC**. Cass (UCNZ). **NC**. Arthurs Pass National Park (LUNZ, NZAC). **MK**. [Mount] Wakefield (FRNZ).**OL**. Hollyford Valley, Hollyford Road end (NZAC). **FD**. Fiordland National Park: Crosscut Range (FRNZ); Doubtful Sound, Deep Cove (LUNZ); Homer Tunnel (FRNZ, LUNZ, NZAC); Lake Manapouri (LUNZ), Wilmot Pass (NZAC); Milford Sound (NZAC); Secretary Island (NZAC). Spey River (LUNZ). Te Anau Downs (NZAC). **SL**. Catlins State Forest Park (LUNZ). MacLennan Range (NZAC). Nugget Point (LUNZ). Slopedown Range (LUNZ). Takitimu Forest (NZAC).

**Stewart Island.** Codfish Island, Sealers Bay (NZAC). Halfmoon Bay (UCNZ). Lee Bay (NZAC). Little Hellfire Beach (LUNZ). Mason Bay (LUNZ). [Mount] Rakeahua hut (LUNZ). Rakeahua [River] valley (NZAC).

**Biology.** *Koroana arthuria* occurs from lowland to higher montane forest margins and shrublands. Collected frequently on *Hebe* species (including *H. odora*), also on *Cassinia* sp., *Coprosma parviflora*, *Metrosideros* sp., *Olearia avicenniaefolia*, and *Brachyglottis buchananii*. Adults collected from November to February throughout the range of the species, with peaks of abundance in late January and February. Apparently univoltine, overwintering as eggs or nymphs; newly emerged adults collected from late November to early December and in January and February. Fore- and hindwings fully developed so probably capable of flight.

Literature record (Myers 1924): Reared in large numbers from nymphs collected beneath stones at Arthurs Pass; small ants were also observed, but myrmecophily was not definitely established; nymphs of this species found in company with those of *Oliarus oppositus,* numerous under stones, in some instances with small ants (*Monomorium* sp.) in the boulder-strewn riverbed at Arthurs Pass (2300 ft [700 m] elevation).

## *Koroana lanceloti* Larivière, 1997

Figures 46, 70; Map 7.

*Koroana lanceloti* Larivière, 1997b: 221.

**Type data. Holotype:** Male (NZAC) labelled "NEW ZEALAND OL; Dart Hut, 945m; 15 Feb 1980; J.C. Watt; beaten at night / HOLOTYPE; *Koroana*; *lanceloti* sp. nov.; Larivière, 1997 (red)." Mounted on card point; genitalia dissected and stored underneath the specimen in genitalia vial containing glycerol.

**Allotype:** Female (NZAC) labelled as holotype.

**Paratypes** (18 males, 22 females) bearing blue paratype labels, with same data as holotype, except as follows: 10 males (5 LUNZ, 5 NZAC), 10 females (4 LUNZ, 6 NZAC), "NEW ZEALAND OL; Dart Hut, 945m; 15 Feb 1980; J.C. Watt; beaten at night"; 2 males (NZAC), 2 females (LUNZ), "NEW ZEALAND OL; Dart Valley, 940m; 17 Feb 1980; J.C. Watt / beaten from; shrubs"; 2 males, 2 females (NZAC), "NEW ZEALAND OL; Dart Hut, 950m; 19 Feb 1980; J.C. Watt / beaten from; *Hebe solicifolia* [sic]; at night"; 4 males (2 LUNZ, 2 NZAC), 7 females (3 LUNZ, 4 NZAC), "NEW ZEALAND OL; Dart Hut; 13–15 Feb 1980; J.S. Dugdale / Malaise trap; in open"; 1 female, "NEW ZEALAND OL; Dart Hut; 13–15 Feb 1980; J.S. Dugdale; pan trap in open".

**Description.** Adult brown, often with a reddish tinge; forewings with dark spots arranged in an irregular transverse band between costa and distal third of clavus.

Vertex brown, with basal compartment often paler; basal emargination deeply V-shaped (more deeply incised than in *K. rufifrons*).

Pronotum pale yellowish brown to brown, often darker laterally. Mesonotum yellowish white, pale brown or, more rarely, mossy green medially, darker brown laterally, often tinged with reddish. Forewings hyaline, sometimes slightly infumate or milky; veins yellowish brown, often nearly black; an irregular transverse band of dark brown spots between costa and apex of clavus, usually darker and better defined than in *K. arthuria*; stigma brown; costa with 17–20 peduncles; Sc+R forked slightly distad of Cu, rarely at same level or basad; r-m slightly distad of $M_{3+4}$, more rarely at same level; A1 and Y-vein often whitish with brown spots near distal third of clavus; apical cells usually 11 (sometimes 10) in number, with 1 or 2 partly or entirely opaque dark brown; tegula slightly darker than pronotum. Legs brown to almost black, with hind tibiae sometimes yellowish brown; fore and mid tibiae annulated blackish and yellowish white; hind tibiae with 3 immovable lateral spines.

Ventral sternites brown to blackish. **Male genitalia**. Anal tube as in *K. arthuria* and *K. rufifrons*. Left genital style as in Fig. 70. Aedeagus (in dorsal view, Fig. 46) with one moderately long (approximately one-third of aedeagal length), sinuate, dorsolateral, spinose process arising apically near the base of the flagellum and 2 thin, short, sickle-shaped, spinose processes visible near the midlength of the periandrium, the dorsally directed process much longer than the less narrowly coiled ventral process.

Body length of males 4.26–5.25 (4.70) mm, of females 4.62–5.63 (5.07) mm.

**Geographical distribution** (Map 7). South Island west coast, from Nelson/Buller to Fiordland, extending eastwards into Otago Lakes area, Central Otago, and the Mount Cook area.

**Material examined.** 324 non-type specimens were seen from the following localities.
**South Island. NN.** Lake Sylvester (LUNZ). Millerton (LUNZ). Mount Arthur Range: Flora Hut/Saddle (NZAC). Westport (LUNZ). **MB.** Rai Valley (FRNZ). Rainbow State Forest, Connors Creek (LUNZ). **BR.** Boatmans Creek (NZAC). Buller Gorge (BPNZ). Capleston (NZAC). Fletchers Creek (NZAC). Lake Moana (CMNZ, LUNZ). Lewis Pass (NZAC). Matakitaki River (LUNZ). Mawhera State Forest (NZAC). Mount Robert (NZAC). Mount Sewell (NZAC). Nelson Lakes National Park: Lake Rotoiti (LUNZ). Punakaiki (LUNZ). Taylorville (NZAC). Tawhai State Forest, Big River road (NZAC). **WD.** Fox Glacier (LUNZ). Franz Josef (LUNZ). Gillespies Beach (NZAC). Hokitika (NZAC). Jackson Bay (NZAC). Lake Paringa (NZAC). Otira (LUNZ), Barrack Creek (NZAC). Poerua River Scenic Reserve (LUNZ). Ross (MONZ). Wanganui River (UCNZ). Westland National Park: Canavans Knob (LUNZ); Lake Mapourika (MONZ); Waiho (NZAC) Gorge (MONZ). **MK.** Hoophorn Stream (NZAC). Mount Cook area (NZAC). **OL.** Mount Aspiring National Park: Arawata Bivouac (LUNZ); Aspiring Hut (LUNZ); Dart Valley/Hut (LUNZ, NZAC); Glacier Burn (LUNZ). Makarora (NZAC). Mount Anstead (NZAC). Queenstown (NZAC). **OL/FD.** Hollyford Valley (LUNZ). **CO.** Kawarau Gorge, Roaring Meg (NZAC). **FD.** Fiordland National Park: Darran Mountains, Middle Gully (NZAC), Tutoko Bench (NZAC); Doubtful Sound (NZAC); Homer Tunnel (NZAC); [Lake Manapouri], Wilmot Pass (NZAC). Milford Sound (NZAC); Secretary Island, on way to Grono Bay (NZAC). Stillwater River (MONZ).

**Biology.** *Koroana lanceloti* is found in lowland to subalpine forest margins and shrublands. Collected regularly on *Olearia* species (including *O. moschata*, *O. avicenniaefolia*, *O. lacunosa*), *Hebe* (especially *H. salicifolia* and *H. subalpina*), and *Coprosma* in these habitats. Other associated plant records include *Carmichaelia*, *Cassinia*, and *Aristotelia fruticosa*. Adults collected from November to April; teneral individuals from November to February and in July. Apparently univoltine, overwintering as eggs or nymphs. Fore- and hindwings fully developed, so probably capable of flight.

**Remarks.** The unbalanced phenotypic expression in male genital characters of individuals from populations in the zones of geographic overlap suggests occasional hybridisation between this species and the others. Hybridisation is suspected in material examined from the following localities: BR - Lake Rotoiti; NN - Mount Arthur Range (*K. lanceloti* x *K. rufifrons*); FD - Doubtful Sound; OL - Hollyford Road end, Homer Tunnel, Milford, Wilmot Pass (*K. lanceloti* x *K. arthuria*).

### *Koroana rufifrons* (Walker) stat. nov.

Figures 18, 45, 69, 107; Map 7.

*Cixius interior* Walker, 1858: 82. Incorrect synonymy of Myers (1927: 689).

*Cixius rufifrons* Walker, 1858: 83.

*Koroana helena* Myers, 1924: 319. Synonymised by Myers (1927: 689).

*Koroana interior* (Walker); –Myers, 1927: 689 (Incorrect combination).

*Koroana interior* (Walker); –Larivière, 1997b: 219.

**Type data. Holotype**: Male (BMNH) labelled "Type (circle with green border) / CIXIUS RUFIFRONS. (One line label folded in two) / N. Zeal.; 54.4". Good condition, double-mounted on card. Abdomen dissected, stored underneath specimen in genitalia vial containing glycerol. [Note: The type locality of specimens collected by Lt-Colonel D. Bolton and bearing the British Museum accession number "54.4" is likely to be Auckland (see Dugdale 1988)].

**Description.** Adult (Fig. 107) yellowish brown, often tinged with reddish orange or green (fading in dead specimens); forewings usually clear, sometimes with a weakly defined pattern of pale brown spots arranged in an irregular transverse band between costa and distal third of clavus.

Vertex ivory to pale yellow, often tinged with brown or reddish orange at middle; basal emargination widely V-shaped.

Pronotum ivory to pale yellowish. Mesonotum ivory to pale yellowish brown medially, darker orange-brown laterally. Forewings hyaline, sometimes slightly infumate or milky; veins yellowish brown, often slightly darker apically; stigma brown, sometimes quite pale; costa with 14–17 peduncles; Sc+R forked distad of Cu, rarely at same level as Cu; r-m usually located at same level as $M_{3+4}$, more rarely slightly distad; A1 and Y-vein yellowish brown; apical cells usually numbering 10 (sometimes 9 or 11), all hyaline; tegula concolorous with pronotum. Legs entirely yellowish or brown, sometimes with base and apex of fore and mid tibiae slightly infuscate; hind tibiae with 3 immovable lateral spines and sometimes a feeble, extra spine present between the 2 basal ones.

Ventral sternite yellowish or brown. **Male genitalia.** Anal tube as in *K. arthuria*. Left genital style as in Fig. 69. Aedeagus (in dorsal view, Fig. 45) with one long (approximately half of aedeagal length), subrectilinear, dorsolateral spinose process located subapically near the base of the flagellum and 2 slender (less robust than in *K. arthuria*), short, sickle-shaped, spinose processes near the midlength of the periandrium, the dorsally directed process slightly longer than the more narrowly coiled ventral process.

Body length of males 4.20–5.40 (4.92) mm, of females 4.90–6.10 (5.38) mm.

**Geographical distribution** (Map 7). Widely distributed in the North Island, with a disjunct distribution in northern and southeastern South Island.

**Material examined.** A total of 335 specimens was seen from the following localities.

**North Island. ND.** Herekino State Forest (NZAC). Hokianga Harbour (NZAC). Kawakawa (NZAC). Mangamuka summit (NZAC). Matarua Forest, Waioku Coach Road (NZAC). Motatau Swamp (AMNZ). Paihia (NZAC). Puketi Forest (NZAC). Tutukaka Harbour (NZAC). Waikaraka Stream (NZAC). Whangape Harbour (NZAC). Waipoua Forest (NZAC). Whangarei (NZAC). **AK.** Auckland (NZAC). Browns Bay (NZAC). Huia (NZAC). Hunua Range (NZAC). Mill Bay (NZAC). Oratia (NZAC). Rangitoto Island (CMNZ). Riverhead Forest (NZAC). Titirangi (LUNZ). Waitakere Range (CMNZ, NZAC). Woodhill (NZAC). **CL.** Egens Park (NZAC). Fourth Branch Scenic Reserve (NZAC). Great Barrier Island, Cliff Island (NZAC). Hikuai Settlement (NZAC). Kauaeranga Valley (LUNZ). Kennedy Block (NZAC). Kirikiri Saddle (NZAC). Little Barrier Island (NZAC). Stony Bay (NZAC). Tairua (NZAC). Tapu (FRNZ, NZAC). Tapu-Coroglen Road (NZAC). Waiaro Bay (NZAC). Waitete Bay (NZAC). **WO.** Mount Pirongia (NZAC). Port Waikato (NZAC). Putaruru (NZAC). Waitomo (NZAC). **BP.** Kaimai-Mamaku Forest Park (NZAC). Lake Rotoma (NZAC). Lake Rotorua (NZAC). Lake Tarawera (LUNZ). Mamaku [Plateau] (FRNZ). Mount Ngongotaha (NZAC). Mount Tarawera (NZAC). Otanga Beach (LUNZ). Rotoehu State Forest (FRNZ). Rotorua (NZAC). Te Aroha (NZAC). Tikitapu/Blue Lake (FRNZ). Urewera National Park, Waimana Valley (NZAC). Waioeka Gorge (NZAC). **TO.** Desert Road (LUNZ). Kaimanawa Forest Park, North (NZAC). Lake Rotoaira (NZAC). Makatote (NZAC). Ohakune (NZAC). Orakeikorako (NZAC). Oturere Stream (NZAC). Pureora State Forest Park (NZAC), Waipapa Reserve (FRNZ, NZAC). Taupo (NZAC). Tihoi (LUNZ). Tongariro National Park: Mount Ruapehu (LUNZ); Tawhai Falls (NZAC). Waikato and Waipakihi Rivers junction (NZAC). Waipunga Falls (NZAC). Whakamaru (NZAC). **HB.** Kaweka Range (NZAC). Puketitiri (NZAC). Putahinu [Ridge] (NZAC). **TK.** Egmont National Park: Taranaki/Mount Egmont (NZAC). Matemateaonga Walkway (NZAC). Ohura (LUNZ). Waitara River, Moki Forest (NZAC). Tangarakau Gorge (NZAC). Wanganui National Park: Whakaroro-Mangapurua track (NZAC). **GB.** Urewera National Park (NZAC): Lake Waikaremoana (MONZ). **RI.** Ruahine Range (NZAC). **WI.** Feilding (NZAC). Manawatu Gorge (BPNZ). **WN.** Keith George Memorial Park (NZAC). Korokoro (NZAC). Ngatiawa River (NZAC). Paekakariki (NZAC). Rimutaka Forest Park (NZAC). Silverstream (NZAC). Tararua Range: Dundas Hut Ridge (NZAC); Mount Holdsworth (MONZ); Waikawa Stream (NZAC). Wellington (NZAC). **WA.** Haurangi State Forest Park, Aorangi Range (NZAC), Ruakokopatuna River tributary (NZAC).

**South Island. SD.** Croisilles Hill (NZAC). D'Urville Island (LUNZ). Port Underwood Saddle (NZAC). Ship Cove (NZAC). Stephens Island (NZAC). Tennyson Inlet (NZAC). Trio Islands (NZAC). **NN.** Anatimo (NZAC). Cobb Reservoir (NZAC). Collingwood (NZAC). Gouland Downs (NZAC). Korere (NZAC). Mangarakau (NZAC). Matai Valley (NZAC). Mount Arthur Range, Flora Hut (NZAC). Mount Burnett (LUNZ). Mount Chrome (NZAC). Nelson (NZAC). Puponga (NZAC). Takaka Hill (NZAC). Whangamoa Saddle (NZAC). **BR.** Nelson Lakes National Park: Lake Rotoiti (NZAC). Tawhai (FRNZ). **MB.** Rainbow State Forest, Chinaman Stream at Wairau River (NZAC). Red Hills (FRNZ). **KA.** Okarahia Stream (CMNZ). **MC.** Cass (NZAC). **SL.** Owaka (NZAC).

**Biology.** *Koroana rufifrons* occurs on trees and shrubs of coastal to submontane forest margins and shrublands, often on stream sides. Often collected on *Hebe parviflora* and other *Hebe* species (including *H. stricta* and *H. divaricata*) and on *Melicytus ramiflorus*. Found less frequently on *Coriaria arborea*, *Fuchsia* and, on rare occasions, on *Hoheria*, *Metrosideros*, *Nothofagus*, *Pittosporum*, *Pseudowintera*, or *Weinmannia*. Large numbers of newly emerged individuals collected in November in the Waimana Valley (Urewera National Park, Bay of Plenty) near a stream at the edge of a mixed podocarp-broadleaf forest by beating mixed vegetation of *Fuchsia-Melicytus-Hebe*. Adults collected from October to April throughout the range of the species with peaks of abundance in January and February. Apparently univoltine, overwintering as eggs or nymphs; newly emerged adults collected in October, November, January, and February. Fore- and hindwings fully developed so probably capable of flight.

## Genus *Malpha* Myers

*Malpha* Myers, 1924: 322. Type species *Malpha muiri* Myers, 1924: 322, by original designation.

**Description.** Brownish yellow species with brown markings on head, thorax, and forewings, especially along costa and on Y-vein.

Vertex approximately 0.6× as long as broad; transverse subapical keel nearly U-shaped, usually connected to anterior margin by 2 short ridges (often obsolescent); basal compartment with a narrow median keel; basal emargination V-shaped. Frons with median carina forked near midlength; median ocellus whitish yellow. Postclypeus with a median carina.

Pronotum with a median longitudinal carina; a pair of curved postocular carinae, one on either side of middle, subparallel or with their midportion reaching close to hind margin. Mesonotum with 5 longitudinal carinae. Forewings about 3× longer than broad; veins with setiferous peduncles, costa with about 20 such peduncles (never more); apical cells 9 in number. Hind tibiae with 3 lateral spines; hind tarsomere I with an apical row of 6–7 teeth, hind tarsomere II with 5 teeth.

**Male genitalia.** Externally visible portions somewhat variable within the genus. Anal tube (Fig. 86) and left genital style (Fig. 71, 72) as illustrated. Aedeagus (Fig. 47, 48, in ventral view) with 3 variously shaped spinose processes arising near the base of the flagellum.

**Remarks.** Endemic genus with 2 species.

## Key to species

1 Frons entirely shining brownish yellow; postclypeus concolorous with frons (Fig. 19). Apex of left genital style of male broadly L-shaped (Fig. 71). Aedeagus (in ventral view, Fig. 47) with 3 long, sinuate processes arising near the base of the flagellum (2 thick processes, one on either side of a much thinner, more acuminate one, the base of which extends along an oblique ridge on the right side of the periandrium) .. ........................ (p.35) ... *Malpha cockcrofti* **Myers**

—Frons with basal portion pale yellowish or sometimes greenish, followed by a wide band of shining dark brown or blackish, next a band of pale yellowish along frontoclypeal suture; postclypeus with base pale yellowish and apex dark brown (Fig. 20). Apex of left genital style of male shaped like the head of a hammer (Fig. 72). Aedeagus (in ventral view, Fig. 48) with 3 subapical spinose processes arising near the base of the flagellum (a long, broad, undulate median process with a tapered end, flanked by 2 shorter, much thinner processes) ................ (p.36) ... *Malpha muiri* **Myers**

## *Malpha cockcrofti* Myers

Figures 19, 47, 71, 108; Map 8.

*Malpha cockcrofti* Myers, 1924: 323.

**Type data. Holotype**: Female (NZAC) labelled "Type (circular red-bordered label) / Otira; 25-12-20; T.C. (handwritten; last line at right angle along right margin) / 161 (handwritten) / J.G. Myers Coll.; B.M. 1937-789. / Holotype; Malpha; cockcrofti; Myers (handwritten; first line at right angle along left border which is red)". Reasonably good condition; costal border of left forewing slightly damaged; left hind tibia glued to point.

**Description.** Adult (Fig. 108) brownish yellow tinged with orange or reddish, marked with chocolate-brown on forewings.

Vertex brownish yellow; transverse subapical keel U-shaped (sometimes narrowly sinuate at middle); basal and anterior compartments nearly equally depressed. Frons (Fig. 19) entirely shining brownish yellow (darker than vertex); frontoclypeal suture subrectilinear; outer carinae less strongly convex at level of frontoclypeal suture than in *M. muiri*. Postclypeus concolorous to frons.

Pronotum very short, paler brownish yellow than head and mesonotum, sometimes almost whitish. Mesonotum brownish yellow, the two inner longitudinal carinae usually paler. Forewings hyaline, clouded with pale orange-brown; costa unmarked; often with a patch of darker brown near base and on disc; veins orange-brown in basal half or alternately brown and whitish, becoming somewhat thickened and continuously dark apically; stigma whitish orange; Sc+R forked basad of Cu; r-m located basad of $M_{3+4}$. Legs brownish yellow tinged with orange (like frons and underneath of thorax).

Ventral sternites brownish. **Male genitalia.** Anal tube as in type species, *Malpha muiri*. Left genital style (Fig. 71) as illustrated. Aedeagus (in ventral view, Fig. 47) with 3 long, sinuate processes arising near the base of the flagellum, 2 thick processes, one on either side of a much thinner, more acuminate one, the base of which extends along an oblique ridge on the right side of periandrium.

Body length of males 3.95, 4.01, 4.06 mm, of females 4.06, 4.39 mm.

**Geographical distribution** (Map 8). South Island west coast.

**Material examined.** A total of 5 specimens was seen from the following localities.
South Island. **BR**. Buller Gorge, Dublin Terrace (NZAC). Fletchers Creek (NZAC). Paparoa Range: adj. Croesus Knob (NZAC); Buckland Peaks (NZAC). **WD**. Otira (NZAC).

**Biology.** Lower mountains to subalpine habitats. Adults collected in November and January. Individuals of both sexes beaten from *Olearia colensoi* (January, adj. Croesus Knob). One specimen recorded from *Celmisia* flowers (November, Buckland Pks). Found in subalpine environment in the South Island (Myers, 1924; type data).

## *Malpha muiri* Myers

Figures 20, 48, 72, 86, 96; Map 8.
*Malpha muiri* Myers, 1924: 322.
*Malpha iris* Myers, 1924: 323. **syn. nov.**

**Type data. Holotype**: Male (NZAC) labelled "Type (circular red-bordered label) / Mt. Alpha; 3600; 14.2.21 (handwritten) / 157 (handwritten) / J.G. Myers Coll.; B.M. 1937-789. / 157 (with pygofer mounted between 2 round plastic pieces) / Holotype; Malpha; muiri; Myers (handwritten; first line at right angle along left border which is red)". Excellent condition; wings spread out; right foretarsi missing.

**Description.** Species slightly paler than *M. cockcrofti*. Adult pale yellowish brown, sometimes tinged with orange or reddish, marked with brown.
Vertex yellow; transverse subapical keel U-shaped (not sinuate at middle); basal compartment depressed; anterior compartment slightly convex, not usually depressed or, if so, never as much as basal compartment. Frons (Fig. 20) with basal portion pale yellowish or sometimes greenish, followed by wide band of shining dark brown or blackish, next a band of pale yellowish along frontoclypeal suture which is subrectilinear; outer carinae strongly convex at level of frontoclypeal suture. Postclypeus with base pale yellowish and apex dark brown.
Pronotum very short, yellowish, sometimes almost whitish. Mesonotum yellowish brown at middle, darker brown, 2 inner longitudinal carinae usually paler. Forewings hyaline, clouded with yellowish white; costa with 3–4 ill-defined brown marks; base and apical third usually fumate; veins pale yellowish brown in basal half or alternately brown and whitish, becoming apically somewhat thickened and continuously dark; stigma whitish brown; Sc+R forked basad of Cu; r-m located basad of $M_{3+4}$. Legs yellowish; fore and mid femora and tibiae with a proximal and distal ring of brown.

Ventral sternites brownish. **Male genitalia.** Pygofer (Fig. 96), anal tube (Fig. 86), and left genital style (Fig. 72) as illustrated. Aedeagus (in ventral view, Fig. 48) with 3 subapical spinose processes arising near the base of the flagellum, a long, broad, undulate median process with a tapered end, flanked by 2 shorter, much thinner processes of unequal length.

Body length of males 3.81, 4.26, 4.31 mm, of females 4.51, 5.29 mm.

**Geographical distribution** (Map 8). Known from only a few North Island and South Island localities.

**Material examined.** A total of 5 specimens was seen from the following localities.
**North Island. WN**. Tararua Range, Mount Alpha (NZAC). York Bay (NZAC, type locality of *M. iris*). **South Island. BR**. Lewis Pass (NZAC).

**Biology.** Adults, including tenerals, collected in November. Taken on the undergrowth of shrubby *Senecio* and *Olearia* in a *Nothofagus* forest (Myers 1924).

**Remarks.** The female holotype of *M. iris* (NZAC) was seen. Characters of the external morphology suggest that it is conspecific with *M. muiri*.

## Genus *Parasemo* gen. nov.

Type species *Parasemo hutchesoni* sp. nov., by present designation.

**Description.** Smallish (4–5 mm), stubby cixiids. General colour brown with a yellow tinge and slightly fumate or milky forewings with a dark spot at tip of each apical cell, and an irregular pattern of brown spots along costa, clavus and on disc.

Vertex approximately 0.6× as long as broad; transverse subapical keel regularly arcuate, not connected to anterior margin by short ridges; basal compartment with an incomplete median keel; basal emargination broadly U-shaped, often notched in middle. Frons (Fig. 21) slightly swollen; median carina absent; frontoclypeal suture rather strongly arcuate; median ocellus present. Postclypeus swollen, without a visible median carina.

Pronotum with a pair of curved postocular carinae, one on either side of middle, subparallel to hind margin. Mesonotum with 5 longitudinal carinae (2 inner ones often evanescent). Forewings about 3× longer than broad; veins with setiferous peduncles, costa with less than 10 such peduncles; apical cells 8 in number. Hind tibiae with 2 robust lateral spines near base; hind tarsomeres I and II, each with an apical row of 7 teeth.

**Male aedeagus** (in ventral view) with 2 robust lateral spinose processes arising near the base of the flagellum.

**Remarks.** Examination of collection material previously identified as *Semo* revealed this new taxon which resembles the latter only superficially. The high dissimilarity in external characters, e.g., the 5 longitudinal carinae of the pronotum, the presence of setiferous peduncles on the forewings, and the configuration of the male genitalia, suggests that it is not congeneric with it. *Semo* species show a high degree of similarity in morphological characters and virtually identical male genitalia suggesting that they form a distinct monophyletic group.

## *Parasemo hutchesoni* sp. nov.

Figures 49, 73, 87, 97, 109.

**Type data. Holotype**: Male (NZAC) labelled "NEW ZEALAND TO; Pureora; Waipapa Res, 570m; 24 Nov 1983; J. Hutcheson/ Malaise trap in; shrublands / HOLOTYPE; *Parasemo*; *hutchesoni*; sp. nov.; Larivière, 1999 (red)". Excellent condition; male genitalia dissected, stored underneath specimen in genitalia vial containing glycerol.

**Paratypes** (3 males, 9 females) bearing blue paratype labels, with same data as holotype except for dates of collection: 1 male (NZAC), 20 Oct 1983; 2 females (MONZ), 3 Nov 1983; 1 male (MONZ), 10 Nov 1983; 1 male, 1 female (AMNZ), 17 Nov 1983; 4 females (NZAC), 1 Dec 1983; 1 female (AMNZ), 22 Dec 1983; 1 female (MONZ), 29 Dec 1983.

**Description.** Adult (Fig. 109) brown with a yellow tinge and slightly fumate or milky forewings with a dark spot at tip of each apical cell and an irregular pattern of brown spots along costa, clavus, and on disc.

Vertex brownish with concolorous margins and carinae. Frons yellowish brown. Postclypeus brownish.

Pronotum brownish yellow. Mesonotum brown. Forewing veins dark brown; stigma well-developed, more or less concolorous with remainder of wing; Sc+R forked slightly basad of Cu; r-m located basad of $M_{3+4}$. Legs yellowish brown, slightly paler than rest of body.

Ventral sternites dark brown, almost black. **Male genitalia**. Pygofer (Fig. 97), anal tube (Fig. 87), and left genital style (Fig. 73) as illustrated. Aedeagus (in ventral view, Fig. 49) with 2 rather long (c. 2/3 of aedeagal length), robust, lateral spinose processes of subequal length (left process arising near the periandrium apex; right process arising more basally).

Body length of males 4.15–4.65 (4.41) mm, of females 4.50 - 5.05 (4.81) mm.

**Geographical distribution.** Known only from the North Island: Pureora State Forest Park (TO) and Mount Te Aroha (BP).

**Material examined.** A total of 17 specimens was seen from the following localities: **North Island. BP.** Mount Te Aroha (BMNH, NZAC). **TO.** Pureora State Forest Park, Waipapa Reserve (AMNZ, MONZ, NZAC).

**Biology.** Unknown.

**Remarks.** This species is named after J. Hutcheson (Forest Research Associates, Rotorua) who found all specimens in the type series.

## Genus *Semo* White

*Semo* White, 1879: 217. Type species *Semo clypeatus* White, 1879: 217, by original designation and monotypy.
**Description.** Rather small (3–6 mm), yellowish-brown to dark brown, stubby cixiids; head, pronotum, and mesonotum with paler margins and carinae; forewings whitish yellow or infumate, often opaque, with a dark spot

at tip of apical cells and irregular, often coalescing patterns of brown spots across midportion and apex.

Vertex 0.3–0.6× as long as broad; transverse subapical keel arcuate, connected or not to anterior margin by 2 narrow ridges (often obsolete); basal compartment with incomplete median keel; basal emargination broadly U-shaped, squarish at middle. Frons slightly swollen; median carina absent or evanescent (reduced to a rather flat elevation) and thickened near midlength; frontoclypeal suture strongly arcuate, sometimes squarish at middle; median ocellus present. Postclypeus swollen (more so in males), without a visible median carina.

Pronotum with a pair of curved postocular carinae, one on either side of middle, subparallel to hind margin. Mesonotum with 3 longitudinal carinae, the median one sometimes evanescent. Forewings 2.5–3.5× longer than broad; venation and markings as in Fig. 12; veins, including costal margin, smooth, lacking visible setiferous peduncles (or peduncles obsolete), often contrastingly more calloused and paler than surrounding area; 7–9 apical cells. Hind tibiae with 3 immovable lateral spines (all 3 equidistant in basal half, or 1 or 2 basally plus 1 or 2 more apically), and sometimes a feeble extra spine between the 2 basal most ones; hind tarsomeres I and II, each with an apical row of 7 teeth.

**Male genitalia.** Externally visible portions similar throughout the genus. Anal tube (Fig. 88), and genital styles (Fig. 74–77) as illustrated. Aedeagus (Fig. 50–53) with 2 spinose processes arising near the base of the flagellum.

**Remarks.** Larivière & Hoch (1998) recently reviewed this genus. The synonymy, type data and slightly modified versions of the key to species and descriptions are included here in order to make this faunal review as comprehensive as possible. Additional information is provided on material examined, geographical distribution, and biology.

Until now the main combination of characters used to diagnose *Semo* from other New Zealand genera has been the frons lacking a median carina and the swollen postclypeus (Deitz & Helmore 1979). These characters are, however, much more variable than previously recognised. Most females studied lack a median carina on the frons but some have a slight flat elevation that is thickened near its middle as in most males. In both sexes the postclypeus varies from convex and slightly swollen to strongly swollen and the postclypeus is on average more swollen in males than in females.

As progress was made in the revision of the whole family, it became clear that *Huttia harrisi* Myers, previously known only from the type, is conspecific with *Semo westlandiae*. Thus *Huttia harrisi* is transferred to *Semo* as a senior synonym of that species (see Remarks section for the genus *Huttia*).

### Key to species (males)

**1** Aedeagus (in ventral view, Fig. 51) with 2 short (approximately 0.4–0.5× aedeagal length), thick, arched spinose processes of subequal length arising near the base of the flagellum ................................................
.... (p. 41) ... ***Semo southlandiae* Larivière & Hoch**

*Distribution (Map 9): central eastern and southeastern South Island.*

—Aedeagus (in ventral view, Fig. 50, 52, 53) with 2 longer (approximately 0.6–0.7× aedeagal length), thinner, sinuate spinose processes of unequal length arising near the base of the flagellum ........................................ 2

**2**(1) Aedeagus (in ventral view, Fig. 50) with right spinose process distinctly shorter than left process; both processes with apices directed outwards .........................
............................. (p. 39) ... ***Semo clypeatus* White**

*Distribution (Map 9): central North Island and northern South Island.*

—Aedeagus (in ventral view, Fig. 52, 53) with right spinose process longer than left process; both processes directed towards the right, or apex of right process nearly hook-shaped ........................................... 3

**3**(2) Aedeagus (in ventral view, Fig. 53) with the apex of the right spinose process sinuate, almost hook-shaped. Apex of left genital style (Fig. 77) broad, with outer edge oblique, rectilinear .............................................
............ (p. 40)... ***Semo harrisi* (Myers) comb. nov.**

*Distribution (Map 9): South Island west coast, western Stewart Island, and eastern Southland.*

—Aedeagus (in ventral view, Fig. 52) with apices of both spinose processes directed towards the right, i.e., apex of right process not sinuate, hook-shaped. Apex of left genital style (Fig. 76) narrower with outer edge arcuate .(p. 41)... ***Semo transinsularis* Larivière & Hoch**

*Distribution (Map 9): southern North Island, south of the Central Volcanic Plateau, and northern South Island, mostly in the west.*

## Semo clypeatus White

Figures 1–9, 11, 12, 22, 50, 74, 88, 98, 110; Map 9.

*Semo clypeatus* White, 1879: 217.

**Type data**. Lectotype (not seen) designated by Deitz in Deitz & Helmore (1979): "A male specimen (total length 4.2 mm, width 1.7 mm; figs 23, 25, 31, 32) in the F.B. White collection, Perth Museum and Art Gallery, Scotland." It bears the labels:" N.Z.; H / LARGE CABINET; CASE 32 / BUCHANAN WHITE COLL[n] ; PERTH MUSEUM; 1979.3.77 / Semo clypeatus / LECTOTYPE ♂; Semo; clypeatus; F.B. White; desig. LL Deitz 1979."

**Description**. Adults (Fig. 110) yellowish brown dorsally; forewings completely infumate or opaque whitish, with an irregular band of brown spots across midportion and a few scattered pale patches distally.

Vertex yellowish brown with yellowish ivory margins and carinae and 3 dark spots (sometimes coalescent) on anterior compartment. Frons (Fig. 22) uniformly yellowish or brownish or with scattered dark spots; outer carinae often pale yellow; median ocellus visible, yellowish. Postclypeus yellowish brown, in some specimens much darker along outer carinae.

Pronotum yellowish brown, often with pale outline. Mesonotum yellowish brown, usually darker medially. Thoracic sterna yellowish brown, always paler than ventral sternites. Forewing venation as in Fig. 6 and 12, sometimes with dark spots along Y-vein and other longitudinal veins; the latter yellowish brown, sometimes calloused, pale yellow or whitish; Sc+R usually forked slightly basad of Cu, more rarely at same level as Cu; r-m usually located at same level as $M_{3+4}$; 8–9 apical cells. Legs yellowish brown to dark brown, often with base and apex of femora paler.

Ventral sternites yellowish brown to dark brown, but in some individuals black. **Male genitalia**. Pygofer (Fig. 98), anal tube (Fig. 88), and left genital style (Fig. 74) as illustrated. Aedeagus (in ventral view, Fig. 50) with 2 long (approximately 0.6–0.7× aedeagal length), thick spinose processes arising subapically near the base of the flagellum (right process slightly shorter than left, their apices curved outwards; flagellum nearly as long as aedeagus, directed towards its base).

Body length of males 3.64–4.72 (4.32) mm, of females 4.92–5.38 (5.14) mm.

**Geographical distribution** (Map 9). Central North Island and northern South Island.

**Material examined**. 168 non-type specimens were seen from the following localities:
**North Island**, along the Taupo-line. **TK**. Egmont National Park: Holly Hut (NZAC); Kapuni Valley (NZAC); South Flank [Mount Egmont] (NZAC); Stony River (NZAC). Egmont National Park, Pouakai Range (NZAC): Ahukawakawa Swamp (NZAC); Pouakai Hump (NZAC); Pouakai Trig (NZAC); SE (NZAC); Tatangi Peak (NZAC). **TO**. Pureora State Forest Park, Mount Pureora summit (NZAC). Tongariro National Park, Mount Ruapehu: Chateau (NZAC);Taranaki Falls (NZAC); Whakapapa (NZAC); Whakapapaiti Hut (NZAC). **GB**. Urewera National Park: Aniwaniwa Falls (NZAC); Lake Waikaremoana (NZAC). **South Island**. **NN**. Cobb Reservoir (NZAC). Lake Sylvester (NZAC). **MB**. Mount Isobel track (UCNZ), Waterfall Stream (UCNZ). Rainbow State Forest, Connors Creek (LUNZ). **BR**. Maruia Springs (UCNZ). Nelson Lakes National Park: Cupola Basin (FRNZ, LUNZ); Lake Rotoiti (LUNZ, NZAC). Lewis Pass (NZAC). Mount Robert (LUNZ). Punakaiki (LUNZ). **NC**. Arthurs Pass National Park: Arthurs Pass, summit (LUNZ); Klondyke Corner (LUNZ). **MC**. Cass (NZAC, UCNZ).

**Biology**. Montane to subalpine shrublands and grasslands, often in the vicinity of streams. Found on the following plants: *Hebe stricta* (tenerals and fully mature adults); *Cassinia vauvilliersii, Coprosma-Olearia* associations, *Dracophyllum longifolium, Hebe-Uncinia* associations, *Nothofagus fusca, Senecio eleagnifolius*, tussocks, and mat plants (fully mature adults). Taken in large numbers in late December on *Halocarpus biformis* in subalpine scrubland at the summit of Mount Pureora (TO). Tenerals found from November to January but mostly in November. Fully mature adults collected from November to end of February but in highest numbers in November and January.

**Remarks**. This species co-occurs with *S. transinsularis* on Mount Ruapehu (North Island central Volcanic Plateau) and although this may be an artefact of sampling, label data suggest that both taxa may be separated altitudinally; *S. clypeatus* having been collected only in sites above 1 000 m. *S. clypeatus, S. transinsularis,* and *S. harrisi* are all found in the northwest Nelson area (NN), but there is no record of their having been collected in the exact same locality.

## Semo harrisi (Myers, 1924) comb. nov.

Figures 53, 77; Map 9.

*Huttia harrisi* Myers, 1924: 322.
*Semo westlandiae* Larivière & Hoch, 1998: 440. **syn. nov.**

**Type data. Holotype**: Female (NZAC) labelled "Type (circular red-bordered label) / West coast; S. Is. N.Z.; Feb. 1923; T. Harris (handwritten; last line at right angle along right margin) / J.G. Myers Coll.; B.M. 1937-789. / Holotype; Huttia; harrisi; Myers (handwritten; first line at right angle along left border which is red)". Reasonably good condition; double-mount; left forewing missing.

**Description**. Adult pale yellowish or brownish dorsally, often with a rusty tinge; forewings opaque brown or whitish brown, often pale in proximal third, with rather large brown patches or heavily mottled with brown across midportion and smaller scattered spots distally.

Vertex pale brown, often with a rusty tinge or thick yellow margins and carinae and in some specimens with 3 dark spots on anterior compartment. Frons yellowish or brownish, often with scattered dark spots; outer carinae often pale yellow; median ocellus visible, yellowish or whitish. Postclypeus yellowish brown to almost black.

Pronotum pale brown (in some individuals with a little black), often with pale outline. Mesonotum yellowish brown to rusty brown, usually darker medially. Thoracic sterna yellowish brown, always paler than ventral sternites. Forewing veins yellowish brown, sometimes calloused, yellowish; Sc+R forked basad of Cu; r-m usually located basad of $M_{3+4}$; 7 or 8 apical cells. Legs yellowish brown to dark brown, often with base and apex of femora pale.

Ventral sternites yellowish brown to dark brown. **Male genitalia**. Anal tube as in *S. clypeatus*. Left genital style as in Fig. 77. Aedeagus (in ventral view, Fig. 53) with 2 long (approximately 0.6–0.7× aedeagal length), thin spinose processes (more slender than in *S. clypeatus* and *S. transinsularis*) arising subapically near the base of the flagellum (left process directed towards the periandrium base; right process longer, sinuate, almost hook-shaped, its apex curved slightly towards the right).

Body length of males 3.48–4.40 (4.05) mm, of females 3.72–5.17 (4.46) mm.

**Geographical distribution** (Map 9). South Island west coast, western Stewart Island, and Owaka, eastern Southland.

**Material examined**. A total of 178 specimens was seen from the following localities:
**South Island. NN.** Denniston (NZAC). Mount Arthur Range, Ellis Basin (UCNZ). Mount Domett (NZAC). **BR.** Mount Sewell, TV station (LUNZ, NZAC). Paparoa Range: Buckland Peaks track (NZAC); Lochnagar Ridge (NZAC); Mount Dewar (NZAC); Mount Priestly (NZAC); Mount Priestly-Mount Dewar basins (NZAC). Hochstetter State Forest, Flagstaff Res. (NZAC). **WD.** Fox Glacier, Chancellor Shelf (LUNZ). Franz Josef (NZAC). Mount Aspiring National Park: Haast Pass, Davis Flat (UCNZ). Mark Range (BPNZ). Mount Tuhua (NZAC). Westland National Park: Alex Knob (NZAC); Castle Rocks Valley (LUNZ). **NC.** Arthurs Pass (NZAC). **MK.** Mount Cook National Park: Ball Hut (LUNZ). **OL.** Mount Aspiring National Park: Liverpool Bivouac (LUNZ). Humbolt Mountains, Route Burn (NZAC). The Key summit (NZAC). **DN.** Swampy Summit (BPNZ). **FD.** Fiordland National Park: Alpine Garden, Homer Tunnel to Milford Rd (NZAC); Darran Mountains, Tutoko Bench (NZAC); Gertrude Valley, Homer Hut (LUNZ); head of Lake Orbell (NZAC), north side (NZAC); Homer Valley [Tunnel] (LUNZ); Homer [Saddle/Tunnel] (BPNZ, NZAC); [Lake] Manapouri, Wolfe Flat (NZAC), Wilmot Pass (NZAC); Murchison Mountains, McKenzie Burn (LUNZ), Plateau Creek (LUNZ); Point Burn Valley, Main Flat (NZAC); Secretary Island (NZAC), ridge towards Mount Grono (NZAC); Takahe Valley (NZAC). **SL.** Owaka (NZAC). Table Hill (NZAC). **Stewart Island.** Codfish Island: Loop track (NZAC), Upper Miro track (NZAC), Valley track (NZAC). Mount Anglem (LUNZ). [Tin Range]: Old Tin Hut (UCNZ).

**Biology**. Montane to subalpine shrublands and grasslands, often in the vicinity of streams; on Stewart Island, apparently in podocarp-broadleaved forest. Found mostly on *Dracophyllum* species, including *D. traversii* and *D. longifolium*, but also on *Coprosma propinqua* (tenerals and adults). In addition, adults recorded on *Hebe* sp., *Olearia ilicifolia*, tussocks, and other, yet undetermined, subalpine plants. Tenerals collected from November to January but more abundantly in December. Adults found from November to February but mostly in December.

**Remarks**. Generally of a paler hue than *S. clypeatus* and *S. transinsularis*, with head more heavily outlined in yellow and, in most individuals, a pale patch resembling a 'shoulder strap' in the proximal third of each forewing.

Collection data suggest that the geographic range of this species is parapatric with that of *S. clypeatus* in the north and *S. southlandiae* in the south. At the local population level there is no record of these species living sympatrically with *S. harrisi*.

The reasoning behind the synonymy of *Semo westlandiae* is discussed in the Remarks section for the genus *Huttia* (p. 28).

## Semo southlandiae Larivière & Hoch

Figures 51, 75; Map 9.

*Semo southlandiae* Larivière & Hoch, 1998: 436.

**Type data. Holotype:** Male (NZAC) labelled "NEW ZEALAND SL; Tower Peak 1000m; Takitimu Range; 30 Jan 1976; L.L. Deitz / sweeping bog / HOLOTYPE; *Semo*; *southlandiae* sp. nov.; Larivière & Hoch, 1998 (red)." Note: male genitalia dissected, stored underneath specimen in genitalia vial containing glycerol. **Allotype:** Female (NZAC) labelled "as holotype / ALLOTYPE; *Semo*; *southlandiae* sp. nov.; Larivière & Hoch, 1998 (red)."
**Paratypes** (4 males 5 females) bearing blue paratype labels, with same data as holotype and allotype: 4 males (2 LUNZ, 2 NZAC) and 5 females (2 LUNZ, 3 NZAC).

**Description.** Adult dark brown dorsally with blackish head and mesonotum; forewings opaque, pale yellowish or whitish brown, often mottled with dark brown.
Vertex brown to almost black with yellowish ivory margins and carinae; anterior compartment solid brown or black. Frons yellowish or brown, mottled with dark brown or uniformly dark brown to black; outer carinae often pale yellow; median ocellus visible, yellowish or whitish. Postclypeus yellowish brown to almost black.
Pronotum brown to black, usually with thick, pale outline. Mesonotum blackish. Thoracic sterna yellowish brown, always paler than ventral sternites. Forewings sometimes with dark spots along Y-vein and other longitudinal veins; veins yellowish brown, sometimes calloused, pale yellow; Sc+R forked at same level as Cu, sometimes slightly basad of Cu; r-m located basad of $M_{3+4}$; 8–9 apical cells. Legs yellowish brown to almost black, often with base and apex of femora pale.
Ventral sternites dark brown to black. **Male genitalia.** Left genital style as in Fig. 75. Aedeagus (in ventral view, Fig. 51) with 2 short (approximately 0.4–0.5× aedeagal length), thick, arched spinose processes subapically near the base of the flagellum (processes subequal in length, directed towards the periandrium base).
Body length of males 3.68–4.36 (4.21) mm, of females 4.20–5.25 (4.80) mm.

**Geographical distribution** (Map 9). Central eastern and southeastern South Island.

**Material examined.** A total of 45 specimens was seen from the following localities:
**South Island. MC.** Porters Pass (NZAC). Staveley (NZAC). **MK.** Mount Cook National Park (LUNZ). **DN.** Berwick State Forest, Meggat Burn (BPNZ). **SL.** Blue Mountains (BPNZ). Mokoreta No. 2 (NZAC). Mount Hedgehope (NZAC). Slopedown Range (NZAC). Takitimu Range: Cheviot [Hills] face (NZAC); Tower Peak (LUNZ, NZAC).

**Biology.** Montane to subalpine shrublands and grasslands, often in the vicinity of streams. Found on *Coprosma-Cassinia-Dracophyllum* associations in tussocks, *Hebe odora*, vegetation surrounding bogs and in a *Nothofagus* forest. Adults collected from December to February but apparently most abundant in January.

**Remarks.** In addition to the specific spine configuration of the aedeagus, the dorsal surface of head and mesonotum is distinctly more blackish than in other species. Females, however, can be paler than males.
The geographic range of this species appears nearly parapatric with that of *S. harrisi* in the south and *S. clypeatus* in the north. Collection records suggest that these taxa do not coexist with *S. southlandiae* in the same local community.

## Semo transinsularis Larivière & Hoch

Figures 52, 76; Map 9.

*Semo transinsularis* Larivière & Hoch, 1998: 438.

**Type data. Holotype:** Male (NZAC) labelled "NEW ZEALAND WN; Tararua Ra; Dundas Hut 1250m; 5 Feb 1985; C.F. Butcher / sweeping / HOLOTYPE; *Semo*; *transinsularis* sp. nov.; Larivière & Hoch, 1998 (red)". Male genitalia dissected, stored underneath specimen in genitalia vial containing glycerol. **Allotype:** Female (NZAC) labelled "as holotype / sweeping; tussock and; fern / ALLOTYPE; *Semo*; *transinsularis* sp. nov.; Larivière & Hoch, 1998 (red)".
**Paratypes (**27 males, 19 females) bearing blue paratype labels, with information as follows: 1 female (MONZ), "NEW ZEALAND WN; Tararua Ra; Dundas Hut 1250m; 5 Feb 1985; C.F. Butcher / sweeping; tussock and; fern"; 1 female (NZAC), " NEW ZEALAND WN; Tararua Ra; Dundas Hut 1250m; 5 Feb 1985; B.A. Holloway / sweeping; tussock & *Olearia*"; 1 male (MONZ), "NEW ZEALAND WN; Tararua Ra; Dundas Hut 1250m; 10 Feb 1985; G.W. Gibbs"; 1 female (MONZ), "NEW ZEALAND WN; Tararua Ra; Dundas Hut 1250m; 10 Feb 1985; B.A. Holloway / on Outside; walls of ; toilet"; 1 male (NZAC), "NEW ZEALAND WN; Tararua Ra; Dundas Hut; 10 Feb 1985; C.F. Butcher / sweeping; near stream"; 2 males (NZAC), " NEW ZEALAND WN; Tararua Ra; Dundas Hut 1250m; 10 Feb 1985; C.F. Butcher / sweeping"; 1

female (MONZ), "NEW ZEALAND WN; Tararua Ra; Dundas Hut, 1250m; 6-13 Feb 1985 / G. Hall; Pan trap 6"; 2 females (NZAC), "NEW ZEALAND WN; Tararua Ra; Dundas Hut; 8 Feb 1985; G.W. Gibbs / sweeping; below; hut"; 2 males (MONZ), 3 females (NZAC), "NEW ZEALAND WN; Tararua Ra; Dundas Hut 1250m; Dec 1984; R.C. Craw / sweeping; *Chionochloa*; at night"; 4 males (NZAC), 2 females (NZAC), "NEW ZEALAND WN; Tararua Ra; Dundas Hut 1250m; 6 Dec 1984; R.C. Craw / sweeping; *Chinochloa*; with *Olearia*; *Dracophyllum*"; 4 females (MONZ), "NEW ZEALAND WN; Tararua Ra; Dundas Hut 1250m; 4 Dec 1984; R.C. Craw / ex; *Olearia*; *lacunosa*"; 1 female (MONZ), "NEW ZEALAND; Tararua Ra; Dundas Hut; 4 Dec 1984; J.S. Dugdale / ex *Hebe*; *akanensis* [sic]"; 2 males (MONZ), "NEW ZEALAND WN; Tararua Ra; Dundas Hut 1250m; 4-5 Dec 1984 / B.G. Bennett &; T.K. Crosby; Malaise trap"; 5 males (NZAC), "NEW ZEALAND; Tararua Ra; Dundas Hut 1250m; 5-6 Dec 1984 / B.G. Bennett &; T.K. Crosby; Malaise trap"; 1 male (MONZ), 1 female (MONZ), "NEW ZEALAND; Tararua Ra; Dundas Hut 1250m; 2-3 Dec 1984 / B.G. Bennett &; T.K. Crosby; Malaise trap"; 4 males (MONZ), "NEW ZEALAND WN; Tararua Ra; Dundas Hut 1250m; Nov 1984; R.C. Craw / Beating"; 1 male (NZAC), "NEW ZEALAND WN; Tararua Ra; Dundas Hut 1250m; 28-29 Nov 1984 / B.G. Bennett &; T.K. Crosby; Malaise trap"; 1 male (NZAC), 1 female (NZAC), "NEW ZEALAND; Tararua Ra; Dundas Hut 1200m; 28 Nov - 6 Dec 1984 / B.G. Bennett &; T.K. Crosby; Pan trap 2"; 3 males (MONZ), 2 females (NZAC), " NEW ZEALAND WN; Tararua Ra; Dundas Hut 1150m; 28 Nov - 6 Dec 1984 / B.G. Bennett &; T.K. Crosby; Pan trap 1". Several paratypes are teneral.

**Description.** Adults pale brown dorsally; forewings infumate or opaque pale brown, marked with dark brown across midportion and along costa or heavily mottled throughout.

Vertex brown, sometimes nearly black, often with 3 dark spots on anterior compartment. Frons yellowish brown with or without scattered dark spots; outer carinae often pale yellow; median ocellus visible, yellowish or whitish. Postclypeus yellowish brown to almost black.

Pronotum brown to blackish, often with pale outline. Mesonotum brown to almost black, usually darker medially. Thoracic sterna yellowish brown, always paler than ventral sternites. Forewing veins yellowish, sometimes calloused; Sc+R forked slightly basad of Cu or at same level as Cu; r-m located basad of $M_{3+4}$; 7 (in some individuals 8 or 9) apical cells. Legs yellowish brown to dark brown, often with base and apex of femora pale.

Ventral sternites brown to blackish. **Male genitalia.** Left genital style as in Fig. 76. Aedeagus (in ventral view, Fig. 52) with 2 long (approximately 0.6-0.7× aedeagal length) spinose processes arising subapically near the base of the flagellum (right process longer than left, with apices curved slightly towards the right; right process more distinctly so).

Body length of males 3.28-4.60 (3.93) mm, of females 3.60-5.17 (4.60) mm.

**Geographical distribution** (Map 9). Southernmost areas of the North Island, south of the central Volcanic Plateau, and northernmost areas of the South Island, mostly in the west.

**Material examined.** A total of 101 specimens was seen from the following localities:

**North Island**, south of the Taupo-line. **TO.** Tongariro National Park: Mount Ruapehu (NZAC). Ohakune (NZAC). **RI.** Ruahine Range: Shuteye Camp (NZAC). **WN.** Tararua Range: Dundas Hut/Ridge (MONZ, NZAC); Logan E Basin (NZAC). **South Island. BR.** Fletchers Creek (NZAC). **NN.** Mount Arthur Range (NZAC): Balloon Hut (NZAC); Flora Track (NZAC).

**Biology.** Montane to subalpine shrublands and grasslands, often in the vicinity of streams. Found mostly on *Chionochloa* sp. (tenerals and adults); also on *Hebe rakaiensis*, *Olearia lacunosa*, and *Nothofagus* sp. (adults). Tenerals found in November, December, and February but more abundantly in November. Fully mature adults collected from November to February but mostly in December and February.

**Remarks.** *Semo transinsularis* co-occurs with *S. clypeatus* on Mount Ruapehu (North Island central Volcanic Plateau) where both species may be separated altitudinally; *S. transinsularis* having been collected only in sites below 1000m. In the Nelson area (NN) the geographic range is almost parapatric with that of *S. clypeatus*. There is no record of the occurrence of both species in the same locality.

## Tribe OECLEINI

### Genus *Tiriteana* Myers

*Tiriteana* Myers, 1924: 325. Type species *Tiriteana clarkei* Myers, 1924: 325, by original designation.

**Description.** Very distinctive genus with subtriangular body outline in dorsal view and forewings almost held flat on abdomen. General colour pale brown to chocolate brown, with a defined pattern of dark bands on forewings.

Vertex approximately 0.6× as long as broad; transverse subapical keel subrectilinear, not connected to anterior margin by short ridges; basal compartment with narrow median keel; basal emargination U-shaped. Frons (Fig. 23) with median carina simple (not forked); maximum width slightly more than twice the width below vertex; outer carinae strongly convex near midlength; frontoclypeal suture almost rectilinear; median ocellus visible, concolorous with surrounding areas. Postclypeus not swollen, with median carina.

Pronotum with median longitudinal carina; a pair of curved postocular carinae, one on either side of middle, their midportion almost reaching hind pronotal margin. Mesonotum with 3 well-defined longitudinal carinae. Forewings about 3× as long as broad; veins with setiferous peduncles, the costa with 15 such peduncles; apical cells 8–9 in number. Hind tibiae without lateral spines; hind tarsomere I with an apical row of 6 teeth, hind tarsomere II with 5 teeth.

**Male aedeagus** (in ventral view) with subapical, lobate processes arising near the base of the flagellum.

**Remarks.** Endemic genus including a single species.

### *Tiriteana clarkei* Myers

Figures 23, 54, 67, 89, 99, 111; Map 10.

*Tiriteana clarkei* Myers, 1924: 325.

**Type data. Holotype**: Female (BMNH) labelled "Type (circular red-bordered label) / Mamaku; 28.12.20 (handwritten) / J.G. Myers Coll.; B.M. 1937-789. / Holotype; Tiriteana; clarkei; ♀ Myers (handwritten; first line at right angle along left border which is red)". Good condition; tibiae and tarsi missing from fore and mid legs.

**Description.** Adult (Fig.111) quite distinctive, pale brown to chocolate-brown with almost clear forewings marked by a dark diagonal band across apex. Vertex pale brownish yellow; basal emargination deeply U-shaped. Frons (Fig. 23) pale to dark brown medially, paler brownish yellow along carinae and transversely near vertex. Postclypeus whitish.

Pronotum pale brownish yellow. Mesonotum yellowish brown to brown. Forewings hyaline, clouded with yellowish; veins yellowish brown; a chocolate brown basal spot in addition to dark diagonal band across apex; stigma yellowish brown; Sc+R forked at same level as Cu; r-m located basad of $M_{3+4}$; tegula concolorous to or slightly darker than pronotum. Legs whitish yellow.

Ventral sternites brown. **Male genitalia**. Pygofer (Fig. 99), anal tube (Fig. 89), and left genital style (Fig. 67) as illustrated. Aedeagus (in ventral view, Fig. 54) with 3 subapical lobate processes arising near the base of the flagellum (2 processes on the right side, almost touching at their bases and another process on the left side).

Body length of males 3.55–4.52 (4.15) mm, of females 3.72–4.81 (4.41) mm.

**Geographical distribution** (Map 10). Restricted to the North Island where it is widely distributed.

**Material examined.** A total of 36 specimens was seen from the following localities.

**North Island. ND.** Russell Forest (NZAC). **AK.** Titirangi (NZAC). Waitakere Range: Opunahu Stream (NZAC); Scenic Drive (NZAC). **CL.** Kauaeranga Valley (FRNZ, NZAC). Little Barrier Island: Awaroa Stream (AMNZ); Haowhenua Stream (AMNZ). **WO.** Onewhero (NZAC). **BP.** Hicks Bay (NZAC). Kaimai-Mamaku Forest Park: Leyland O'Brien tramline track (NZAC); Tuahu track (NZAC). Mount Te Aroha, Tui Mine (NZAC). Papatea (NZAC). Rotorua, Galaxy Road (NZAC). Tarukenga [Stream] (NZAC). Tikitapu/Blue Lake (NZAC). Urewera National Park, Waimana Valley (NZAC). **TK.** Egmont National Park: Dawson Falls (NZAC). Matemateaonga Walkway, Kohi Saddle to Omaru Hut (NZAC). **TO.** Pureora State Forest Park, North Block, Okahukura Road (NZAC). **GB.** Kakanui (NZAC). **WN.** Tiritea (NZAC).

**Biology.** Adults collected from October to February but most abundantly in January. Found mainly in or at the edge of broadleaf forests, e.g., *Beilschmiedia tarairi* forests. Recorded from *Coprosma rhamnoides*, *Coprosma* spp., and *Carpodetus serratus* (in large numbers). Apparently univoltine. Numerous tenerals collected in late December on *Carpodetus serratus* (Wanganui National Park, TK). Dispersal power: Adults fully winged; collection in Malaise traps providing indirect evidence of flight.

## Tribe PENTASTIRINI

### Genus *Oliarus* Stål

*Oliarus* Stål, 1862: 306. Type species *Cixius walkeri* Stål, 1859: 272, by subsequent designation of Distant (1906: 256).

**Description** (New Zealand). Greyish or brownish species with head, thorax, and abdomen deep glossy black.
Vertex 0.9–1.7× as long as broad; transverse subapical keel U- or V-shaped, not connected to anterior margin by short ridges; basal compartment without median keel; basal emargination V-shaped. Frons with median carina entire (forked only at a short distance from vertex); outer carinae slightly sinuate, rather convex above frontoclypeal suture; median ocellus visible, yellowish. Postclypeus not swollen, with a median carina.
Pronotum with median longitudinal carina; a pair of curved postocular carinae, one on either side of middle, subparallel or not to hind margin. Mesonotum with 5 sharply defined longitudinal carinae. Forewings about 3× longer than broad; veins with setiferous peduncles, costa with less than 10 such peduncles; apical cells 11 in number. Hind tibiae with 2–3 lateral spines; hind tarsomere I with an apical row of 7–8 teeth, hind tarsomere II with 13 teeth.
**Male genitalia** (New Zealand taxa). Externally visible portions uniform throughout the genus. Anal tube (Fig. 90) and left genital style (Fig. 78–79) as illustrated. Aedeagus (in lateral view, Fig. 55) with 3 short spinose processes arising on the periandrium; flagellum with one process near its base and 2 at its tip.

**Remarks**. Cosmopolitan genus with 2 species endemic to New Zealand. Additional information on the extensive synonymy of this widespread genus is available in (Metcalf 1936).

### Key to species

1 Greyish, with head, thorax, and abdomen deep glossy black. Vertex of head narrow, 1.7× longer than broad (Fig. 26). Maximum width of face, across eyes, more than 4× its width near vertex (Fig. 24). Postclypeus black, with well-developed tawny median carina (Fig. 24). Costa of each forewing with 3–4 irregularly spaced setiferous peduncles. Left genital style of male as in Fig. 78. Larger species, average body length of males 7.19 mm, of females 8.73 mm .................................
................ (p. 44) ... *Oliarus atkinsoni* **Myers**

—Yellowish brown, with head, thorax, and abdomen deep glossy black. Vertex of head about as long as broad (Fig. 27). Maximum width of face, across eyes, less than 3× its width near vertex (Fig. 25). Postclypeus black, with incomplete, shorter tawny median carina (Fig. 25). Costa of each forewing with 6–8 irregularly spaced setiferous peduncles. Left genital style of male as in Fig. 79. Smaller species, average body length of males 5.08 mm, of females 6.03 mm ....................
................ (p. 45) ... *Oliarus oppositus* (**Walker**)

### *Oliarus atkinsoni* Myers

Figures 24, 26, 55, 78, 90, 100, 112; Map 11.

*Oliarus atkinsoni* Myers, 1924: 325.

**Type data. Holotype**: Male (BMNH) labelled "Type (circular red-bordered label) / Gollan's Valley; 5.11.21 ( small handwritten label) / 151 [or 156?] (small handwritten label turned upside down) / J.G. Myers Coll.; B.M. 1937-789. / Holotype; Oliarus; atkinsoni; ♂ Myers (handwritten; first line printed at right angle along left border which is red)". Poor condition; left forewing and abdomen missing as well as tibiae and tarsi from the left fore leg and both hind legs.

**Description.** Adult (Fig. 112) greyish, with head, thorax, and abdomen deep glossy black.
Vertex (Fig. 26) deep glossy black with yellow borders, approximately 1.7× longer than broad; transverse subapical keel U-shaped; basal emargination deeply V-shaped. Frons (Fig. 24) deep glossy black with tawny carinae; maximum width across eyes more than 4× width near vertex; frontoclypeal suture regularly U-shaped (more narrowly so than in *O. oppositus*). Postclypeus black with a well-defined median carina, almost entirely tawny.
Pronotum deep glossy black with a pair of curved, postocular carinae, one on either side of middle, their inner portion reaching close to hind margin. Forewings slightly more than 3× longer than broad, hyaline, clouded with whitish; veins pale yellowish brown; costa unmarked, with 3–4 irregularly spaced setiferous peduncles; Sc+R forked at about same level as Cu; r-m located basad of $M_{3+4}$; stigma brownish; tegula yellowish to blackish. Legs blackish, except for edges of segments and most of hind tibiae which are tawny or pale yellow, with 2–3 close-set lateral spines (in basal half); hind tarsomere I with an apical row of 8 teeth, hind tarsomere II with 13 teeth.
Ventral sternites black. **Male genitalia**: Pygofer (Fig. 100), anal tube (Fig. 90), and left genital style (Fig. 78) as

illustrated. Aedeagus (in lateral view, Fig. 55) with 3 stout spinose processes on the periandrium (median process longer than lateral ones, and more acuminate); flagellum with 2 long apical spinose processes and a shorter basal one.

Body length of males 6.42–7.83 (7.19) mm, of females 7.75–9.25 (8.73) mm.

**Geographical distribution** (Map 11). General in the North Island wherever its host (*Phormium*) occurs; also recorded from one South Island locality in Buller.

**Material examined.** A total of 101 specimens was seen from the following localities.
**North Island. ND.** North Cape, Spirits Bay (NZAC). **AK.** Mauku Stream (NZAC). Waitakere Range, Te Henga (NZAC). **CL.** Little Barrier Island, Maraeroa (MONZ). **BP.** Kaimai-Mamaku Forest Park (NZAC). **TO.** Lake Taupo (NZAC). Owhango (UCNZ). Turangi, Frethey Drive (NZAC). **TK.** Paiaka (NZAC).**WN.** Gollans Valley (NZAC). Palliser Bay (MONZ). Waikanae (NZAC). **South Island. BR.** Springs Junction (south end of Palmers Road) (NZAC).

**Biology.** Occurs in flax swamps (*Phormium* sp.) where it can be found, often in copula, on the shaded side of *Phormium* leaves. Adults collected from November to March. Reported to have a two-year life cycle, most of which is occupied by the nymphal stages (Cumber 1952b). For additional information on life cycle, biology, ecology, distribution, and the role of this species as vector of the Yellow Leaf Disease of *Phormium,* see Cumber (1952a–c; 1953a, b; 1954a–d).

## *Oliarus oppositus* (Walker)

Figures 25, 27, 79; Map 11.

*Cixius oppositus* Walker, 1851: 345.

*Oliarus oppositus* (Walker); –White, 1879: 216.

*Cixius marginalis* Walker, 1858. Synonymised by Myers, 1927: 690.

**Type data. Holotype**: Male (BMNH) labelled " Type (circular green-bordered label) / nZeal (circular, handwritten) / 27. CIXIUS OPPOSITUS, (one line label folded in two)". Very good condition; double-mounted on piece of white rubber-like material.

**Description.** A much smaller species than *O. atkinsoni*. Adult yellowish brown, with head, thorax, and abdomen deep glossy black.

Vertex (Fig. 27) deep glossy black with yellow borders, approximately 0.9× as long as broad; transverse subapical keel shallowly V-shaped; basal emargination shallowly V-shaped. Frons (Fig. 25) deep glossy black with yellow carinae; maximum width across eyes less than 3× width near vertex; frontoclypeal suture squarish, more widely U-shaped than in *O. atkinsoni*. Postclypeus black with a median carina, yellowish for a short distance from frontoclypeal suture.

Pronotum deep glossy black with a pair of curved postocular carinae, one on each side of middle, subparallel to hind margin. Forewings nearly 3× longer than broad, hyaline, faintly suffused with greyish yellow; costa unmarked; veins yellowish brown; apical crossveins sometimes darker; costa with 6–8 irregularly spaced setiferous peduncles; Sc+R forked distad of Cu; r-m located basad of $M_{3+4}$; stigma brownish; tegula yellowish. Legs blackish, except for edges of segments and most of hind tibiae which are tawny, with 3 lateral spines; hind tarsomere I with an apical row of 7 teeth, hind tarsomere II with 13 teeth.

Ventral sternites black. **Male genitalia**. Anal tube and aedeagus similar to *O. atkinsoni*. Left genital style as in Fig. 79.

Body length of males 3.00–5.67 (5.08) mm, of females 5.42–6.67 (6.03) mm.

**Geographical distribution** (Map 11). Extremely common throughout New Zealand.

**Material examined.** A total of 2074 specimens was seen from over 250 localities in the following areas.
**Three Kings Islands. North Island**. ND, AK, CL, WO, BP, TK, TO, HB, GB, RI, WI, WN, WA. **South Island**. SD, NN, MB, KA, BR, WD, NC, MC, SC, MK, OL, CO, DN, FD, SL. **Stewart Island.**

**Biology**. Lowland to subalpine environments. Occurs in natural as well as modified habitats, on low herbage (especially grasses) rather than bushes like most other New Zealand Cixiidae. Adults collected from October to April but mostly during the summer months. Apparently univoltine.

**Remarks.** Individuals of subalpine to alpine populations are usually darker in colour and smaller in size.

# REFERENCES

Crosby, T.K.; Dugdale, J.S.; Watt, J.C. 1998: Area codes for recording specimen localities in the New Zealand subregion. *New Zealand Journal of Zoology 25*: 175–183.

Cumber, R.A. 1952a: Entomological aspects of Yellow-Leaf Disease of *Phormium*. *New Zealand Science Review 10*: 3–4.

—— 1952b: Studies on *Oliarus atkinsoni* Myers (Hem.: Cixiidae), vector of the "Yellow-Leaf" Disease of *Phormium tenax* Forst. I. - Habits and environment, with a note on natural enemies. *New Zealand Journal of Science and Technology, Section B, 34*: 92–98.

—— 1952c: Studies on *Oliarus atkinsoni* Myers (Hem.: Cixiidae), vector of the "Yellow-Leaf" Disease of *Phormium tenax* Forst. II. - The nymphal instars and seasonal changes in the composition of nymphal populations. *New Zealand Journal of Science and Technology, Section B, 34*: 160–165.

——1953a: The New Zealand species of *Oliarus* (Hem. Cixiidae). *Transactions of the Royal Society of New Zealand 81*: 71–72.

—— 1953b: Studies on *Oliarus atkinsoni* Myers (Hem.: Cixiidae), vector of the "Yellow-Leaf" Disease of *Phormium tenax* Forst. III. - Resistance of nymphal forms to submergence-control by inundation. *New Zealand Journal of Science and Technology, Section B, 34*: 260–266.

—— 1953c: Investigations into the Yellow-Leaf Disease of *Phormium*. IV. - Experimental induction of Yellow-Leaf condition in *Phormium tenax* Forst. by the insect vector *Oliarus atkinsoni* Myers. (Hem., Cixiidae). *New Zealand Journal of Science and Technology, Section A, 34*: 31–40.

—— 1954a: Die-back condition of *Phormium* seedlings used in Yellow-Leaf investigations. *New Zealand Journal of Science and Technology, Section A, 35*: 270–272.

—— 1954b: Search for alternative vectors of the Yellow-Leaf disease of *Phormium*. *New Zealand Journal of Science and Technology, Section A, 36*: 32–37.

—— 1954c: Studies on *Oliarus atkinsoni* Myers (Hemiptera: Cixiidae), vector of the "Yellow-Leaf" Disease of *Phormium tenax* Forst. IV. - Disease-vector relationships. *New Zealand Journal of Science and Technology, Section B, 35*: 530–549.

—— 1954d: Injury to *Phormium* caused by insects, mites, and molluscs. *New Zealand Journal of Science and Technology, Section A, 36*: 60–74.

Curtis, J. 1837: Cixius. *British Entomology 14*: pl. 673.

Deitz, L.L.; Helmore, D.W. 1979: Illustrated key to the families and genera of planthoppers (Homoptera: Fulgoroidea) from the New Zealand sub-region. *New Zealand Entomologist 7(1)*: 11–19.

Distant, W.L. 1906: Rhynchota. The fauna of British India, including Ceylon and Burma. No. 3. Taylor & Francis, London, England: 1–266.

Dugdale, J.S. 1988: Lepidoptera — annotated catalogue and keys to family-group taxa. *Fauna of New Zealand 14*. 262 pp.

Emeljanov, A.F. 1971: New USSR genera of Cixiidae and Issidae (Homoptera, Auchenorrhyncha). [In Russian.] *Entomologicheskoye Obozreniye 50(3)*: 619–627. English translation: *Entomological Review 50(3)*: 350–354.

—— 1989: On the problem of the division of the family Cixiidae (Homoptera, Cicadina). [In Russian] *Entomologicheskoye Obozreniye 68(1)*: 93–106. English translation: *Entomological Review 68(4)*: 54–67.

—— 1997: The ways of developing classification and reconstructing phylogeny in the family Cixiidae. *Program & Abstract Book, 9th International Auchenorrhyncha Congress*, Sydney, 17–21 February 1997: 38–39.

Fennah, R.G. 1975: New cavernicolous cixiid from New Zealand (Homoptera: Fulgoroidea). *New Zealand Journal of Zoology 2(3)*: 377–380.

Kramer, J.P. 1981: Taxonomic study of the planthopper genus *Cixius* in the United States and Mexico (Homoptera: Fulgoroidea: Cixiidae). *Transactions of the American Entomological Society 107 (1–2)*: 1–68.

Kuschel, G. (Ed.) 1975: Biogeography and ecology in New Zealand. W. Junk Publishers, The Hague. 689 pp.

Larivière, M.-C. 1995: Cydnidae, Acanthosomatidae, and Pentatomidae (Insecta: Heteroptera): systematics, geographical distribution, and bioecology. *Fauna of New Zealand 35*. 112 pp.

—— 1997a: New Zealand Cixiidae (Hemiptera): Taxonomy, faunal composition, regional diversity, and ecological preferences. *Program & Abstract Book, 9th International Auchenorrhyncha Congress*, Sydney, 17–21 February 1997: 25.

—— 1997b: Taxonomic review of *Koroana* Myers (Hemiptera: Cixiidae), with description of a new species. *New Zealand Journal of Zoology 24*: 213–223.

——; Hoch, H. 1998: The New Zealand planthopper genus *Semo* White (Hemiptera: Cixiidae): taxonomic review, geographical distribution, and biology. *New Zealand Journal of Zoology 25*: 429–442.

Latreille, P.A. 1804: Histoire naturelle, générale et particulière des crustacés et des insectes 12: 1–424. Dufart, Paris.

Linnaeus, C. 1758: Systema Naturae per regna triae naturae, secundum classes, ordines, genera, species, cum characteribus, differentiis, synonymis, locis. Editio decima, reformata: i-v, 1–824. Salvii, Holmiae.

Metcalf, Z.P. 1936: General catalogue of the Hemiptera. Fascicle IV, Fulgoroidea. Part 2, Cixiidae. Smith College, Northampton: 3–269.

—— 1938: The Fulgoroidea of Barro Colorado and other parts of Panama. *Bulletin of the Museum of Comparative Zoology 83(5)*: 277–243 [sic].

Millar, I. 1998: Beneath the trees. *Newsletter of the Project Crimson Trust (*Winter 1998). 2 pp.

Muir, F. A. G. 1922: New Malayan Cixiidae (Homoptera). *Philippine Journal of Science 20(1)*: 11–119.

—— 1923: On the classification of the Fulgoroidea (Homoptera). *Proceedings of the Hawaiian Entomological Society 5*: 205–247.

—— 1925: On the genera of Cixiidae, Meenoplidae and Kinnaridae (Fulgoroidea, Homoptera). *Pan-Pacific Entomologist 1(3)*: 97–110; *1 (4)*: 156-163.

—— 1931: Descriptions and records of Fulgoroidea from Australia and the South Pacific Islands. No. I. *Records of the Australian Museum 18*: 63–83.

Myers, J.G. 1924: The New Zealand plant-hoppers of the family Cixiidae (Homoptera). *Transactions and Proceedings of the New Zealand Institute 55*: 15–26.

—— 1927: On the nomenclature of New Zealand Homoptera. *Transactions and Proceedings of the New Zealand Institute 57*: 685–690.

O'Brien, L.B.; Wilson, S.W. 1985: Planthopper systematics and external morphology. Pp. 61–102 In: Nault, L.R.; Rodriguez, J.G. (*eds*). The Leafhoppers and Planthoppers. New York, John Wiley & Sons. 500 pp.

Sheppard, C.; Martin, P.G.; Mead, F.W. 1979. A planthopper (Homoptera: Cixiidae) associated with red imported fire ant (Hymenoptera: Formicidae) mounds. *Journal of the Georgia Entomological Society 14*: 140–144.

Stål, C. 1859: Hemiptera species novas descripsit. Pp. 219–298 In: *Kongliga svenska Fregatten Eugenies Resa omkring jorden, under Befäl af C.A. Virgin aren 1851–1853.* Hemiptera. Norstedt: Stockholm, vol. III, Zoologi, Insekter.

—— 1862: Novae el minus cognitae Homopterorum formae et species. *Berliner Entomologische Zeitschrift 6*: 303–315.

Van Stalle, J. 1991: Taxonomy of the Indo-Malayan Pentastirini (Homoptera, Cixiidae). *Bulletin de l'Institut royal des Sciences naturelles de Belgique, Entomologie 61*: 5–101.

Walker, A.K.; Crosby, T.K. 1988: The preparation and curation of insects (revised edition). *New Zealand Department of Scientific and Industrial Research information series 163* : 91 p.

Walker, F. 1851: List of the specimens of Homopterous insects in the collection of the British Museum 2: 261–636. British Museum, London.

—— 1858: List of the specimens of homopterous insects in the collection of the British Museum. Supplement. British Museum, London. 1–307.

Wardle, P. 1991: Vegetation of New Zealand. Cambridge University Press, Cambridge. 672 p.

White, F.B. 1879: List of the Hemiptera of New Zealand. *The Entomologist's Monthly Magazine 15*: 213–220.

Wilson, S.W.; Mitter, C.; Denno, R.F.; Wilson, M.R. 1994. Evolutionary patterns of host plant use by delphacid planthoppers and their relatives. Pp. 7–113 *in* Denno, R.F.; Perfect, T.J. (eds). Planthoppers. Their ecology and management. New York, Chapman & Hall. 799 pp.

Wise, K.A. 1977: A synonymic checklist of the Hexapoda of the New Zealand sub-region. *Bulletin of the Auckland Institute and Museum 11*: 1–176.

Zimmerman, E.C. 1948: Homoptera: Auchenorrhyncha. *Insects of Hawaii 4:* 1–268.

**Appendix A.** Geographical coordinates of main localities. Coordinates should read as 00°00'S/ 000°00'E. The two-letter area codes follow Crosby *et al.* (1998).

Alderman Islands, CL .................................. 3658/17605
Alpine Garden, FD ...................................... 4445/16757
Anatimo, NN ............................................... 4049/17256
Aorangi I., ND ............................................. 3529/17444
Arawata Bivouac, WD ................................. 4425/16836
Aspiring Hut, WD ........................................ 4429/16840
Aniwaniwa Falls, GB .................................. 3844/17711
Arawata River, WD ..................................... 4400/16840
Arthur's Pass, NC ....................................... 4255/17133
Arthur's Pass National Park, NC ................. 4254/17141
Auckland, AK .............................................. 3651/17446
Awatotara Forest, CH ................................. 4402/17636

Bauza Island, FD ........................................ 4518/16655
Berwick State Forest, DN ........................... 4559/17000
Blue Mountains, SL .................................... 4554/16923
Bluff Hill, SL ................................................ 4636/16820
Boatmans Creek, BR .................................. 4203/17154
Browns Bay, AK .......................................... 3643/17445
Buckland Peaks, BR ................................... 4153/17138
Buller Gorge, BR ......................................... 4151/17143

Canavans Knob, WD ................................... 4323/17010
Canister Cove Scientific Reserve, CH ........ 4420/17613
Capleston, BR ............................................. 4204/17150
Cass, MC .................................................... 4302/17145
Castaway Camp, Great Island, TH ............. 3410/17208
Castle Rocks Valley, WD ............................ 4327/17009
Catlins State Forest Park, SL ..................... 4620/16900
Chatham Island, CH ................................... 4400/17630
Chetwode Islands, SD ................................ 4054/17405
Cheviot Hills, SL ......................................... 4238/16745
Chinaman Stream, MB ................................ 4150/17256
Christchurch, MC ........................................ 4330/17245
Christmas Village, SI .................................. 4645/16759
Cliff Island, CL ............................................ 3610/17527
Cobb Reservoir, NN .................................... 4107/17240
Codfish Island, SI ....................................... 4647/16738
Collingwood, NN ......................................... 4041/17241
Coromandel, CL .......................................... 3645/17530
Council Cave, NN ....................................... 4052/17250
Croesus Knob, BR ...................................... 4218/17123
Croisilles Hill, SD ........................................ 4106/17335

Crosscut Range, FD ................................... 4446/16802
Cupola Basin, BR ........................................ 4159/17245
Cuvier Island, CL ........................................ 3626/17546

Darran Mountains, FD ................................ 4439/16801
Dart Hut/Valley, OL ..................................... 4431/16834
Davis Flat, OL ............................................. 4408/16919
Dawson Falls, TK ........................................ 3919/17406
Days Bay, WN ............................................. 4117/17454
Dean's Bush, MC ........................................ 4332/17236
Denniston, NN ............................................ 4144/17148
Desert Road, TO ........................................ 3914/17544
Dolamore Park, SL ..................................... 4604/16849
Doubtful Sound, FD .................................... 4516/16651
Douglas Creek, WD .................................... 4357/16921
Dundas Ridge/Hut, WN .............................. 4043/17528
Dunedin, DN ............................................... 4553/17030
Dun Mountain, NN ...................................... 4121/17322
D'Urville Island, SD .................................... 4050/17351

Egens Park, CL ........................................... 3651/17534
Egmont National Park, TK .......................... 3917/17404
Eves Valley, NN .......................................... 4120/17304

Fantail Creek, CL ........................................ 3631/17520
Feilding, WI ................................................. 4013/17534
Fiordland National Park, FD ....................... 4553/16657
Fletchers Creek, BR ................................... 4159/17150
Fourth Branch Scenic Reserve, CL ............ 3709/17545
Fox Glacier, WD .......................................... 4330/17007
Franz Josef, WD ......................................... 4325/17010
Freds Camp, SI ........................................... 4656/16759

Gertrude Valley, FD .................................... 4445/16801
Gillespies Beach, WD ................................. 4325/16949
Glacier Burn, WD ........................................ 4427/16849
Glory Bay, Pitt Island, CH ........................... 4419/17612
Glory Scenic Reserve, Pitt Island, CH ....... 4419/17612
Gollans Valley, WN ..................................... 4120/17453
Gouland Downs, NN ................................... 4054/17219
Great Barrier Island, CL ............................. 3613/17524
Great Island, TH ......................................... 3410/17208
Grebe Valley, FD ......................................... 4535/16722
Greymouth, BR ........................................... 4227/17112

Halfmoon Bay, SI ........................................ 4654/16809
Haast, WD ................................................... 4353/16903
Haurangi State Forest Park, WA ................ 4131/17519

| | | | |
|---|---|---|---|
| Hen & Chicken Islands, ND | 3556/17444 | Lake Rotorua, BP | 3805/17617 |
| Herekino State Forest, ND | 3513/17313 | Lake Sylvester, NN | 4106/17238 |
| Hicks Bay, BP | 3735/17817 | Lake Tarawera, BP | 3813/17624 |
| Hikuai Settlement, CL | 3704/17546 | Lake Taupo, TO | 3855/17549 |
| Hochstetter State Forest, BR | 4224/17135 | Lake Waikaremoana, GB | 3846/17705 |
| Hokianga Harbour, ND | 3531/17322 | Lee Bay, SI | 4651/16807 |
| Hokitika, WD | 4243/17058 | Lewis Pass, BR | 4223/17224 |
| Hokonui Hills, SL | 4606/16843 | Little Barrier Island, CL | 3612/17505 |
| Holly Hut, TK | 3916/17403 | Little Hellfire Beach, SI | 4652/16745 |
| Hollyford Valley, OL | 4449/16807 | Liverpool Bivouac, OL | 4426/16840 |
| Homer Tunnel, FD | 4446/16759 | Lochnagar Ridge, BR | 4204/17133 |
| Hoophorn Stream, MK | 4346/17005 | Lottin Point Road, BP | 3732/17810 |
| Huia, AK | 3700/17434 | Longwood Range, SL | 4613/16750 |
| Humboldt Mountains, OL | 4444/16815 | Lynfield, AK | 3656/17443 |
| Hunua Range, AK | 3705/17512 | | |
| | | MacLennan Range, SL | 4631/16918 |
| Jackson Bay, WD | 4358/16837 | Makarora, OL | 4414/16914 |
| | | Makatote, TO | 3916/17521 |
| Kaihoka Lakes, NN | 4033/17236 | Mamaku, BP | 3806/17605 |
| Kaimai-Mamaku Forest Park, BP | 3752/17556 | Mamaku Plateau, BP | 3803/17604 |
| Kaimanawa Forest Park/Range, TO | 3857/17610 | Manawatu Gorge, WI | 4018/17547 |
| Kaingaroa, Pitt Island, CH | 4414/17615 | Mangahuia Stream, RI | 3955/17555 |
| Kaitaia, ND | 3507/17316 | Mangamuka, ND | 3513/17333 |
| Kakanui, GB | 3739/17824 | Mangamuka Hill, ND | 3512/17330 |
| Kapuni Valley, TK | 3920/17407 | Mangarakau, NN | 4039/17229 |
| Kauaeranga Valley, CL | 3708/17537 | Mark Range, WD | 4357/16907 |
| Kawakawa, ND | 3508/17404 | Marlborough State Forest, ND | 3539/17342 |
| Kawarau Gorge, CO | 4502/16908 | Maruia Springs, BR | 4223/17220 |
| Kaweka Forest Park/Range, HB | 3917/17622 | Mason Bay, SI | 4655/16745 |
| Keith Gorge Memorial Park, WN | 4106/17505 | Matai Valley, NN | 4116/17317 |
| Kennedy Bay, CL | 3641/17534 | Matakitaki River, BR | 4213/17231 |
| Kennedy Block, CL | 3639/17530 | Matarua Forest, ND | 3534/17336 |
| Kerikeri, ND | 3514/17357 | Matemateaonga Walkway, TK | 3917/17445 |
| Kirikiri Saddle, CL | 3710/17539 | Mauku Stream, AK | 3711/17448 |
| Klondyke Corner, NC | 4300/17135 | Mawhera State Forest, BR | 4228/17130 |
| Korere, NN | 4132/17248 | Mercury Islands, CL | 3638/17552 |
| Korokoro, WN | 4113/17452 | Milford Sound, FD | 4441/16756 |
| | | Mill Bay, AK | 3700/17436 |
| Lake Kaniere, WD | 4250/17109 | Millerton, NN | 4138/17153 |
| Lake Manapouri, FD | 4531/16727 | Moana, BR | 4235/17129 |
| Lake Mapourika, WD | 4319/17013 | Mokoreta, SL | 4625/16904 |
| Lake Matheson, WD | 4327/16958 | Motatau Swamp, ND | 3530/17401 |
| Lake Moana, BR | 4237/17127 | Motuhoropapa I., AK | 3641/17457 |
| Lake Orbell, FD | 4517/16740 | Motuti River, ND | 3523/17326 |
| Lake Paringa, WD | 4343/16925 | Mount Alpha, WN | 4059/17516 |
| Lake Rotoaira, BP | 3901/17542 | Mount Anglem, SI | 4644/16755 |
| Lake Rotoiti, BR | 4150/17250 | Mount Anstead, OL | 4430/16837 |
| Lake Rotoma, BP | 3803/17635 | Mount Arthur Range, NN | 4113/17241 |

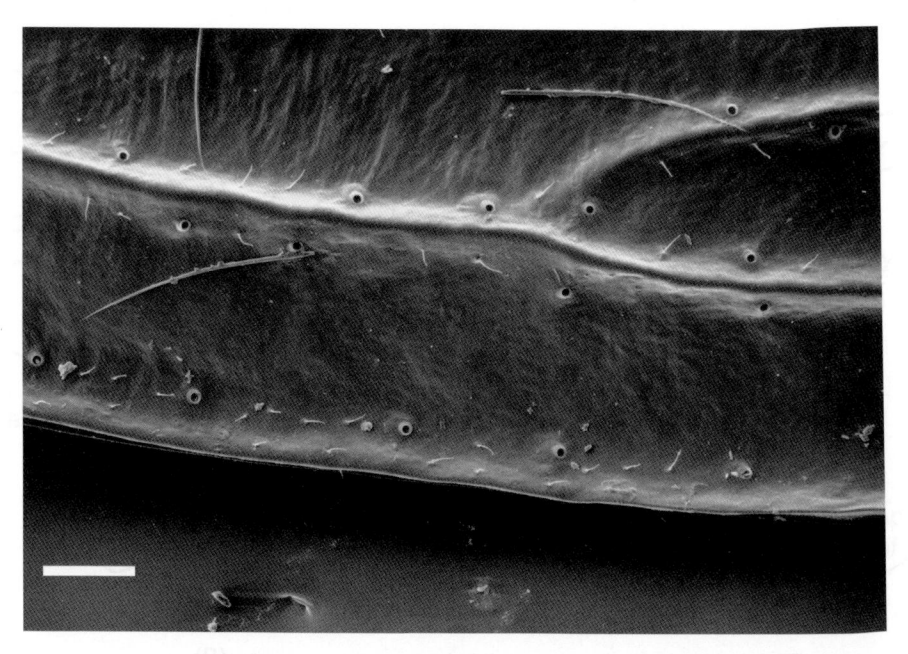

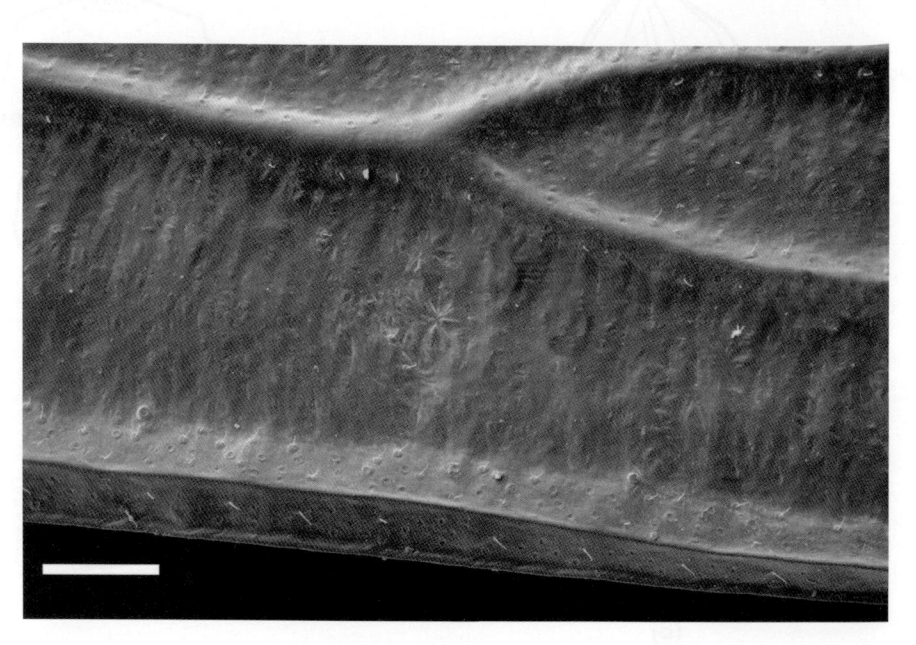

**Fig. 10–11** Scanning electron micrographs showing part of the forewings: (10) *Aka finitima*, veins with setiferous peduncles; (11) *Semo clypeatus*, veins without setiferous peduncles. Scale bar = 0.1 mm.

Hen & Chicken Islands, ND ......................... 3556/17444
Herekino State Forest, ND .......................... 3513/17313
Hicks Bay, BP.............................................. 3735/17817
Hikuai Settlement, CL ................................. 3704/17546
Hochstetter State Forest, BR ....................... 4224/17135
Hokianga Harbour, ND ............................... 3531/17322
Hokitika, WD .............................................. 4243/17058
Hokonui Hills, SL ....................................... 4606/16843
Holly Hut, TK.............................................. 3916/17403
Hollyford Valley, OL ................................... 4449/16807
Homer Tunnel, FD ....................................... 4446/16759
Hoophorn Stream, MK ................................. 4346/17005
Huia, AK ..................................................... 3700/17434
Humboldt Mountains, OL ............................ 4444/16815
Hunua Range, AK ........................................ 3705/17512

Jackson Bay, WD ......................................... 4358/16837

Kaihoka Lakes, NN ...................................... 4033/17236
Kaimai-Mamaku Forest Park, BP ................ 3752/17556
Kaimanawa Forest Park/Range, TO ............. 3857/17610
Kaingaroa, Pitt Island, CH ........................... 4414/17615
Kaitaia, ND ................................................. 3507/17316
Kakanui, GB ............................................... 3739/17824
Kapuni Valley, TK ....................................... 3920/17407
Kauaeranga Valley, CL ................................ 3708/17537
Kawakawa, ND ........................................... 3508/17404
Kawarau Gorge, CO .................................... 4502/16908
Kaweka Forest Park/Range, HB .................. 3917/17622
Keith Gorge Memorial Park, WN ................ 4106/17505
Kennedy Bay, CL ........................................ 3641/17534
Kennedy Block, CL ..................................... 3639/17530
Kerikeri, ND ............................................... 3514/17357
Kirikiri Saddle, CL ...................................... 3710/17539
Klondyke Corner, NC .................................. 4300/17135
Korere, NN ................................................. 4132/17248
Korokoro, WN ............................................ 4113/17452

Lake Kaniere, WD ....................................... 4250/17109
Lake Manapouri, FD ................................... 4531/16727
Lake Mapourika, WD .................................. 4319/17013
Lake Matheson, WD ................................... 4327/16958
Lake Moana, BR ......................................... 4237/17127
Lake Orbell, FD .......................................... 4517/16740
Lake Paringa, WD ....................................... 4343/16925
Lake Rotoaira, BP ....................................... 3901/17542
Lake Rotoiti, BR ......................................... 4150/17250
Lake Rotoma, BP......................................... 3803/17635

Lake Rotorua, BP ........................................ 3805/17617
Lake Sylvester, NN ..................................... 4106/17238
Lake Tarawera, BP ...................................... 3813/17624
Lake Taupo, TO .......................................... 3855/17549
Lake Waikaremoana, GB ............................ 3846/17705
Lee Bay, SI ................................................. 4651/16807
Lewis Pass, BR ........................................... 4223/17224
Little Barrier Island, CL ............................. 3612/17505
Little Hellfire Beach, SI ............................. 4652/16745
Liverpool Bivouac, OL ............................... 4426/16840
Lochnagar Ridge, BR .................................. 4204/17133
Lottin Point Road, BP ................................ 3732/17810
Longwood Range, SL ................................. 4613/16750
Lynfield, AK ............................................... 3656/17443

MacLennan Range, SL ............................... 4631/16918
Makarora, OL ............................................. 4414/16914
Makatote, TO ............................................. 3916/17521
Mamaku, BP ............................................... 3806/17605
Mamaku Plateau, BP .................................. 3803/17604
Manawatu Gorge, WI ................................. 4018/17547
Mangahuia Stream, RI ................................ 3955/17555
Mangamuka, ND ........................................ 3513/17333
Mangamuka Hill, ND ................................. 3512/17330
Mangarakau, NN ........................................ 4039/17229
Mark Range, WD ........................................ 4357/16907
Marlborough State Forest, ND .................... 3539/17342
Maruia Springs, BR .................................... 4223/17220
Mason Bay, SI ............................................ 4655/16745
Matai Valley, NN ........................................ 4116/17317
Matakitaki River, BR .................................. 4213/17231
Matarua Forest, ND .................................... 3534/17336
Matemateaonga Walkway, TK ..................... 3917/17445
Mauku Stream, AK ..................................... 3711/17448
Mawhera State Forest, BR .......................... 4228/17130
Mercury Islands, CL ................................... 3638/17552
Milford Sound, FD ..................................... 4441/16756
Mill Bay, AK .............................................. 3700/17436
Millerton, NN ............................................. 4138/17153
Moana, BR.................................................. 4235/17129
Mokoreta, SL .............................................. 4625/16904
Motatau Swamp, ND .................................. 3530/17401
Motuhoropapa I., AK .................................. 3641/17457
Motuti River, ND ........................................ 3523/17326
Mount Alpha, WN ...................................... 4059/17516
Mount Anglem, SI ...................................... 4644/16755
Mount Anstead, OL .................................... 4430/16837
Mount Arthur Range, NN ........................... 4113/17241

Mount Aspiring National Park, OL ............. 4409/16909
Mount Burnett, NN ...................................... 4038/17238
Mount Camel Peninsula, ND ...................... 3449/17310
Mount Cook, MK ......................................... 4336/17009
Mount Cook National Park, MK ................. 4337/17010
Mount Chrome, NN ..................................... 4142/17302
Mount Dewar, BR ........................................ 4205/17133
Mount Domett, NN ...................................... 4104/17219
Mount Hedgehope, SL ................................. 4606/16843
Mount Hercules, WD ................................... 4310/17027
Mount Holdsworth, WN ............................... 4052/17525
Mount Isobel, MB ........................................ 4229/17251
Mount Manaia, ND ...................................... 3549/17432
Mount Ngongotaha, BP ............................... 3807/17612
Mount Orowhana, ND ................................. 3514/17314
Mount Pirongia, WO ................................... 3759/17506
Mount Priestly, BR ...................................... 4204/17132
Mount Pureora, TO ...................................... 3833/17538
Mount Rakaehua, SI .................................... 4657/16753
Mount Robert, BR ........................................ 4150/17249
Mount Ruapehu, TO .................................... 3918/17534
Mount Sewell, BR ........................................ 4224/17121
Mount Stokes, SD ........................................ 4103/17406
Mount Tarawera, BP .................................... 3813/17631
Mount Te Aroha, BP .................................... 3732/17545
Mount Tuhua, WD ....................................... 4249/17111
Mount Wakefield, MK ................................. 4342/17007
Murchison Mountains, FD .......................... 4515/16732

Nelson, NN .................................................. 4117/17317
Nelson Lakes National Park, BR ................. 4156/17241
New River, BR ............................................. 4233/17108
Ngatiawa River, WN ................................... 4054/17506
Noises Islands, AK ...................................... 3642/17458
North Cape, ND ........................................... 3425/17303
Nugget Point, SL .......................................... 4627/16948

Oaro, KA ...................................................... 4231/17330
Ohakune, TO ................................................ 3925/17525
Ohura, TK .................................................... 3851/17459
Okarahia Stream, KA .................................. 4234/17330
Okiwi Bay, SD ............................................. 4106/17340
Okuru, WD ................................................... 4354/16855
Omahuta State Forest, ND .......................... 3514/17338
Onewhero, WO ............................................ 3719/17455
Opononi, ND ................................................ 3530/17324
Orakeikorako, TO ........................................ 3829/17609

Oratia, AK ................................................... 3655/17437
Orete Forest, BP .......................................... 3738/17757
Otaki River Fork, WN ................................ 4052/17514
Otanga Beach, BP ....................................... 3733/17809
Otira, WD .................................................... 4250/17134
Oturere Stream, TO ..................................... 3911/17547
Owaka, SL ................................................... 4627/16940
Owhango, TO .............................................. 3900/17522

Paekakariki, WN .......................................... 4059/17457
Paiaka, TK ................................................... 3910/17436
Paihia, ND ................................................... 3517/17405
Palliser Bay, WN/WA ................................. 4126/17504
Palmerston North, WI ................................. 4022/17537
Paparoa Range, BR ..................................... 4205/17133
Papatea, BP ................................................. 3740/17751
Pitt Island, CH ............................................. 4417/17612
Poerua River Scenic Reserve, WD ............. 4309/17032
Point Burn Valley, FD ................................ 4518/16739
Poor Knights Islands, ND ........................... 3528/17444
Porarari River, BR ...................................... 4205/17121
Port Pegasus, SI .......................................... 4713/17640
Port Underwood Saddle, SD ....................... 4118/17407
Port Waikato, WO ....................................... 3724/17444
Port William, SI .......................................... 4650/16805
Porters Pass, MC ......................................... 4318/17145
Pouakai Range, TK ..................................... 3915/17401
Puhi Puhi Valley, KA .................................. 4219/17343
Puketi Forest, ND ....................................... 3514/17346
Puketitiri, HB .............................................. 3917/17632
Punakaiki, BR .............................................. 4207/17120
Puponga, NN ............................................... 4031/17243
Pureora State Forest Park, TO .................... 3832/17537
Putaihinu Ridge, HB ................................... 3837/17704
Putaruru, WO .............................................. 3804/17547

Queenstown, OL ......................................... 4502/16840

Rainbow State Forest, MB ......................... 4156/17258
Rai Valley, NN ............................................ 4114/17335
Rakeahua River ........................................... 4659/16753
Rangitoto Island, AK .................................. 3648/17452
Red Hills, MB ............................................. 4144/17301
Red Island, CL ............................................ 3638/17556
Rereauira, BP .............................................. 3735/17804
Riccarton Bush, MC .................................... 4332/17236
Rimutaka Forest Park/Range, WN ............. 4115/17502
Riverhead, AK ............................................ 3645/17436

Roaring Billy Forest Walk, WD .................. 4356/16918
Ross, WD ...................................................... 4254/17048
Ross Creek Reservoir, DN .......................... 4551/17030
Rotoehu State Forest, BP ........................... 3755/17631
Rotorua, BP ................................................. 3809/17615
Route Burn, OL ........................................... 4445/16815
Ruahine Forest Park/Range, RI .................. 4004/17603
Ruamahuanui I., CL .................................... 3657/17606
Russell Forest, ND ...................................... 3516/17417

Secretary Island, FD ................................... 4514/16656
Ship Cove, SD ............................................. 4106/17414
Silverstream, WN ........................................ 4119/17441
Slopedown Range, SL ................................. 4622/16904
Spey River, FD ............................................ 4533/16713
Spirits Bay, ND ............................................ 3426/17248
Stanley Island, CL ....................................... 3639/17553
Stillwater River, FD ..................................... 4504/16719
Springs Junction, BR ................................... 4220/17211
Staveley, MC ............................................... 4339/17126
Stephens Island, SD .................................... 4040/17400
Stony Bay, CL .............................................. 3631/17525
Stony River, TK ........................................... 3916/17358
Swampy Summit, DN .................................. 4548/17028

Table Hill, SL .............................................. 4604/16955
Tairua, CL .................................................... 3700/17551
Takahe Valley, FD ....................................... 4517/16740
Takaka, NN .................................................. 4051/17248
Takaka Hill, NN ........................................... 4102/17251
Takitimu Forest/Range, SL ......................... 4543/16751
Tangarakau Gorge, TK ................................ 3859/17451
Tangihua Range, ND ................................... 3553/17407
Tapu, CL ...................................................... 3659/17530
Tapu-Coroglen Road, CL ............................ 3659/17538
Tapu Hill, CL ............................................... 3659/17530
Taranaki/Mount Egmont, TK ...................... 3918/17440
Tararua Forest Park/Range, WN ................. 4103/17520
Tarukenga, BP ............................................. 3805/17609
Taupo, TO .................................................... 3841/17605
Tawhai, BR .................................................. 4209/17146
Tawhai Falls, TO ......................................... 3910/17530
Tawhai State Forest, BR ............................. 4210/17152
Tawhiti Rahi, ND ......................................... 3528/17444
Taylorville, BR ............................................ 4226/17119
Te Anau Downs, FD .................................... 4511/16750
Te Aroha, WO/BP ........................................ 3732/17543
Te Paki Coastal Reserve, ND ...................... 3428/17248
Te Paki Trig, ND ......................................... 3428/17246
Tennyson Inlet, SD ...................................... 4107/17346
The Key, OL/SL ........................................... 4533/16754
Three Kings Islands, TH ............................. 3410/17207
Tihoi, TO ..................................................... 3837/17537
Tikitapu/Blue Lake, BP ............................... 3812/17620

Tin Range, SI .............................................. 4705/16745
Tiritea, WN .................................................. 4025/17540
Titirangi, AK ............................................... 3656/17440
Tongariro National Park, TO ...................... 3909/17538
Tower Peak, SL ........................................... 4539/16748
Trio Islands, SD ........................................... 4050/17400
Turangi, TO ................................................. 3900/17549
Tutoko River, FD ........................................ 4440/16800
Tutukaka Harbour, ND ................................ 3537/17432

Upper Hutt, WN .......................................... 4107/17504
Urewera National Park, TO/GB .................. 3830/17700

Waiaroho, BP ............................................... 3736/17808
Waiheke Island, AK .................................... 3648/17508
Waiho Gorge, WD ....................................... 4322/17010
Waikanae, WN ............................................. 4053/17504
Waikaraka Stream, ND ................................ 3517/17345
Waikare River, ND ...................................... 3521/17413
Waikato-Waipakihi Rivers junction, TO ..... 3914/17547
Waikawa Stream, WN ................................. 4041/17509
Waikawau, CL ............................................. 3636/17531
Waimana Valley, BP ................................... 3804/17700
Waioeka Gorge, BP ..................................... 3815/17717
Waiomu, CL ................................................ 3702/17531
Waipapa Reserve, TO .................................. 3818/17541
Waipaua-Glory Bay, Pitt Island, CH ........... 4418/17612
Waipaua Scenic Reserve, Pitt Island, CH ... 4418/17612
Waipoua State Forest, ND ........................... 3537/17332
Waipunga Falls, TO ..................................... 3857/17632
Waitakere Range, AK .................................. 3659/17432
Waitangi, Chatham Island, CH .................... 4357/17633
Waitara River, TK ....................................... 3859/17444
Waitete Bay, CL .......................................... 3639/17526
Waitomo, WO .............................................. 3814/17507
Wanganui, WI .............................................. 3957/17503
Wanganui River, WD .................................. 4303/17025
Warawara State Forest, ND ......................... 3523/17319
Wellington, WN ........................................... 4115/17445
Westland National Park, WD ...................... 4332/17005
Westport, NN ............................................... 4145/17136
Whakamaru, TO .......................................... 3826/17548
Whakaroro-Mangapurua track, TK ............. 3908/17503
Whangamoa Saddle, NN ............................. 3909/17442
Whangape Harbour, ND .............................. 3523/17313
Whangarei, ND ............................................ 3543/17419
Whangarei Heads, ND ................................. 3552/17432
Whatupuke Island, ND ................................ 3554/17445
Wilmot Pass, FD .......................................... 4531/16711
Wolfe Flat, FD ............................................. 4532/16717
Woodhill, AK ............................................... 3645/17426

York Bay, WN ............................................. 4116/17454

**Appendix B.** Native plants associated with Cixiidae species.

*Agathis australis* (D. Don.) Loudon ............................................... Araucariaceae
*Aristotelia fruticosa* Hook. f. ........................................................ Elaeocarpaceae
*Ascarina lucida* Hook. f. .............................................................. Chloranthaceae
*Beilschmiedia tarairi* (A. Cunn.) Benth. et Hook. f. ex Kirk ........... Lauraceae
*Blechnum capense* (L.) Schlecht. .................................................. Blechnaceae
*Blechnum* sp. ................................................................................ Blechnaceae
*Brachyglottis buchananii* (J.B. Armst.) B. Nordenstam ................... Asteraceae
*Brachyglottis huntii* (F. Muell.) B. Nord. ....................................... Asteraceae
*Carmichaelia* sp. ............................................................................ Fabaceae
*Carpodetus serratus* J.R. Forst. et G. Forst. ................................... Escalloniaceae
*Cassinia* sp. .................................................................................. Asteraceae
*C. vauvilliersii* (Homb. et Jacq.) Hook. f. ...................................... Asteraceae
*Chionochloa* sp. ............................................................................ Poaceae
*Coprosma* sp. ................................................................................ Rubiaceae
*C. chathamica* Cockayne .............................................................. Rubiaceae
*C. parviflora* Hook. f. ................................................................... Rubiaceae
*C. propinqua* A. Cunn. ................................................................. Rubiaceae
*C. rhamnoides* A. Cunn. ............................................................... Rubiaceae
*Coriaria arborea* Lindsay ............................................................. Coriariaceae
*Dacrydium cupressinum* Lamb. .................................................... Podocarpaceae
*Dracophyllum* sp. ......................................................................... Epacridaceae
*D. longifolium* (J.R. et G. Forst.) R. Br. ....................................... Epacridaceae
*D. subulatum* Hook. f. .................................................................. Epacridaceae
*Fuchsia* sp. ................................................................................... Onagraceae
*Halocarpus biformis* (Hook.) Quinn ............................................. Podocarpaceae
*Halocarpus kirkii* (Parl.) Quinn ................................................... Podocarpaceae
*Hebe* sp. ....................................................................................... Scrophulariaceae
*H. divaricata* (Cheesem.) Cockayne & Allan ............................... Scrophulariaceae
*H. odora* (Hook. f.) Cockayne ..................................................... Scrophulariaceae
*H. parviflora* (Vahl) Cockayne & Allan ....................................... Scrophulariaceae
*H. rakaiensis* (J.B. Armst.) Ckn. .................................................. Scrophulariaceae
*H. salicifolia* (G. Forst.) Pennell .................................................. Scrophulariaceae
*H. stricta* (Benth.) L.B. Moore ..................................................... Scrophulariaceae
*H. subalpina* (Cockayne) Cockayne & Allan ................................ Scrophulariaceae
*Hoheria* sp. ................................................................................... Malvaceae
*Melicytus* sp. ................................................................................ Violaceae
*M. chathamicus* (F. Muell.) Garn.-Jones ...................................... Violaceae
*M. ramiflorus* J.R.Forst & G. Forst. ............................................. Violaceae
*Metrosideros* sp. ........................................................................... Myrtaceae
*Myoporum* sp. ............................................................................... Myoporaceae
*Nothofagus* sp. .............................................................................. Fagaceae
*N. fusca* (Hook. f.) Oersted .......................................................... Fagaceae
*Olearia* sp. ................................................................................... Asteraceae
*O. avicenniaefolia* (Raoul) Hook. f. ............................................. Asteraceae
*O. colensoi* Hook. f. ..................................................................... Asteraceae
*O. ilicifolia* Hook. f. ..................................................................... Asteraceae
*O. lacunosa* Hook. f. .................................................................... Asteraceae
*O. moschata* Hook. f. ................................................................... Asteraceae
*Phormium* sp. ............................................................................... Phormiaceae
*Pittosporum* sp. ............................................................................ Pittosporaceae
*Prumnopitys ferruginea* (D. Don.) de Laub .................................. Podocarpaceae
*Pseudowintera* sp. ........................................................................ Winteraceae
*Pseudopanax* sp. .......................................................................... Araliaceae
*P. crassifolius* (Sol. ex A. Cunn.) C. Koch ................................... Araliaceae
*P. simplex* (= *Neopanax simplex*) (G. Forst.) Philipson ............. Araliaceae
*Schefflera digitata* J.R. Forst. et G. Forst .................................... Araliaceae
*Senecio eleagnifolius* Hook. f. ..................................................... Asteraceae
*Solanum aviculare* var. *albiflorum* Cheesem. ............................. Solanaceae
*Uncinia* sp. ................................................................................... Cyperaceae
*Weinmannia* sp. ............................................................................ Cunoniaceae
*W. racemosa* Linn. f. .................................................................... Cunoniaceae
*Xeronema* sp. ............................................................................... Phormiaceae

# ILLUSTRATIONS

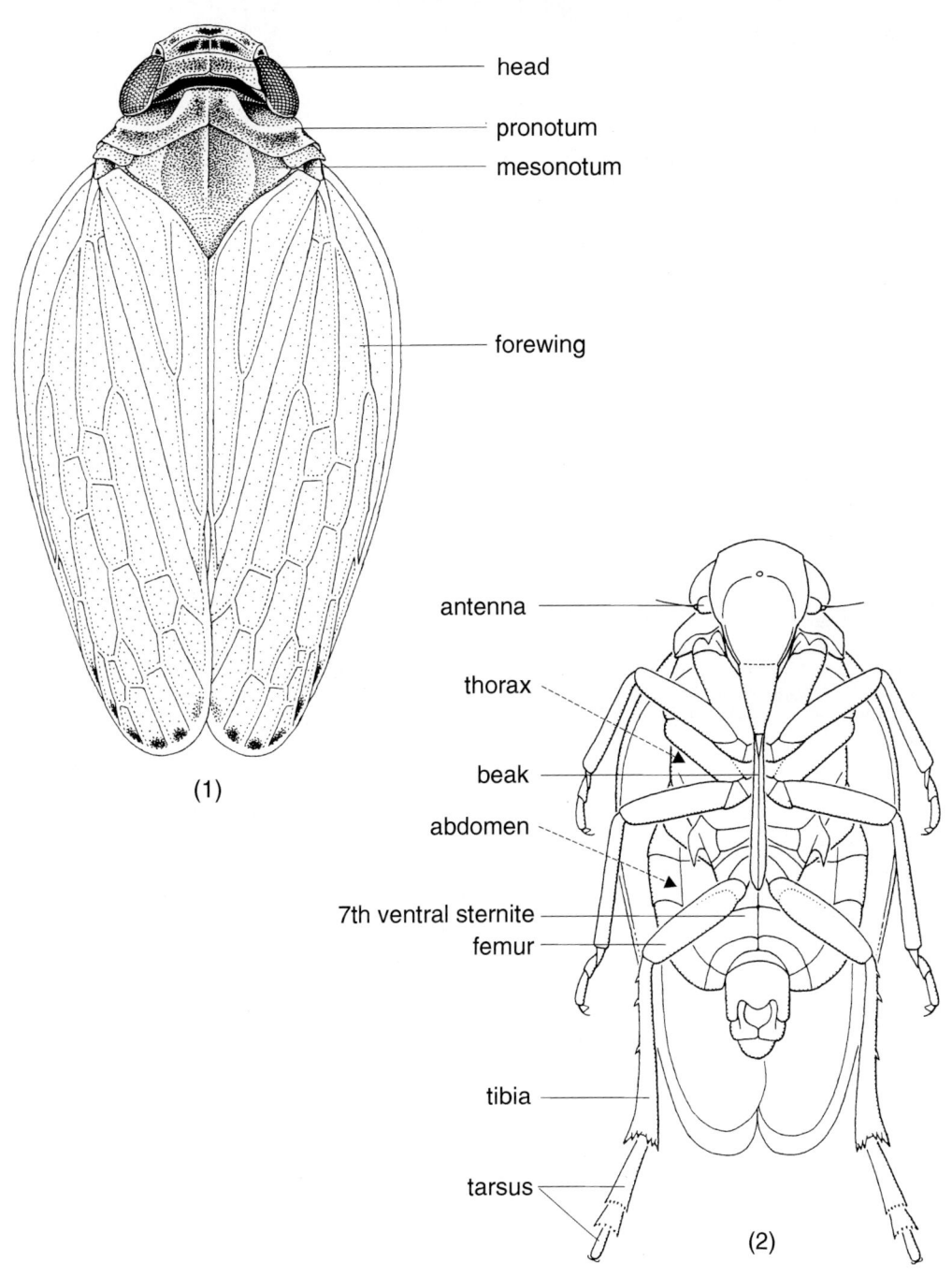

**Fig. 1, 2** External morphology of *Semo clypeatus*: (1) female, dorsal; (2) male, ventral.

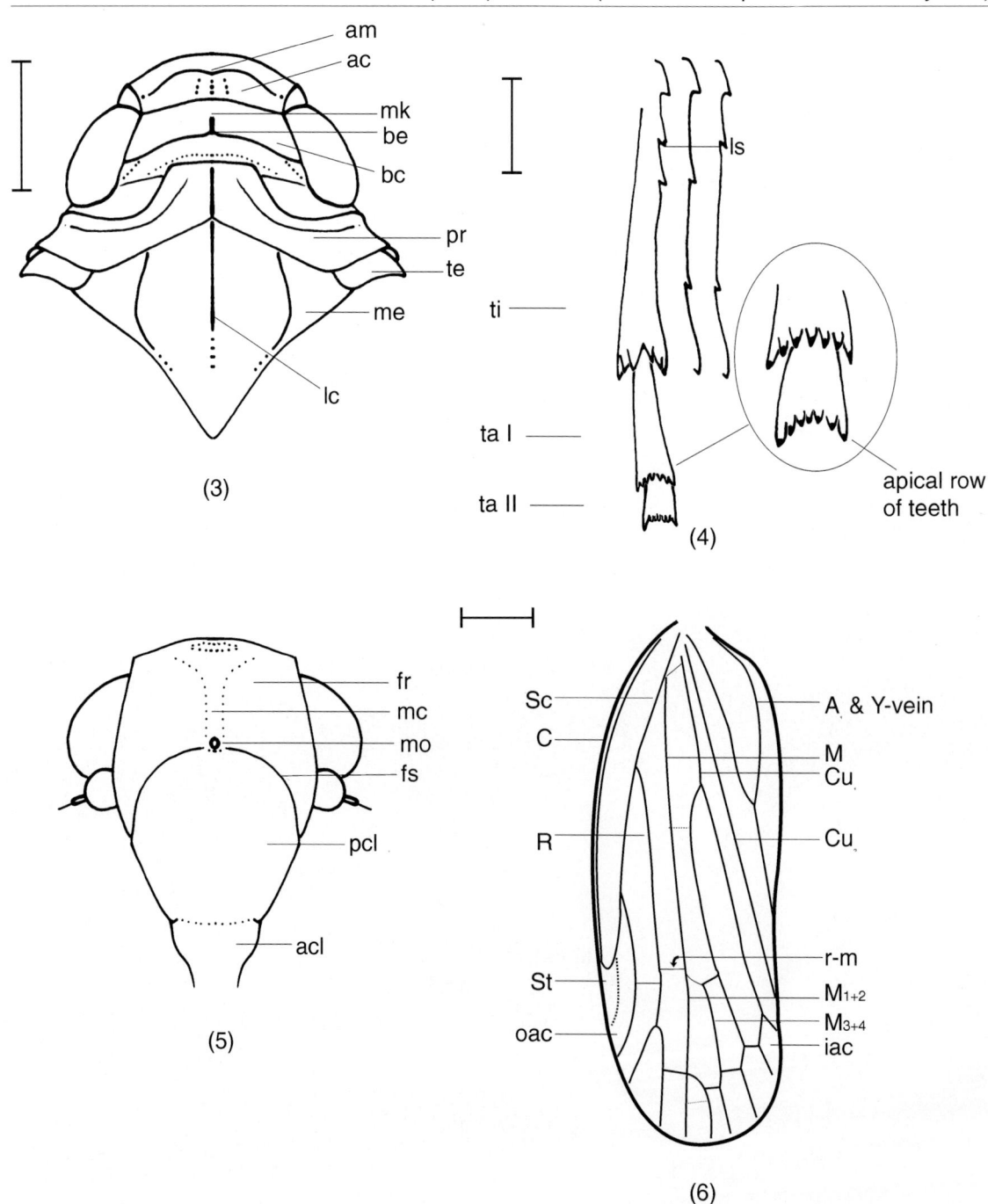

**Fig. 3–6** Diagnostic features of the external morphology, *Semo clypeatus*: (3) head, pronotum, and mesonotum, dorsal view (ac – apical compartment of vertex; am – anterior margin of vertex; bc – basal compartment of vertex; be – basal emargination of vertex; lc – longitudinal carina; me – mesonotum; mk – median keel; pr – pronotum; te – tegula); (4) left hind leg, ventral view (ls – lateral spines; ta I – tarsomere I; ta II – tarsomere II; ti – tibia); (5) head, frontal view (acl – anteclypeus; fr – frons; fs – frontoclypeal suture; mc – median carina; mo – median ocellus; pcl – postclypeus); (6) left forewing (A1 – first anal vein and Y-vein; C – costa; Cu – cubital vein; iac – inner apical cell; M – median vein; oac – outer apical cell; r-m – crossvein between R and M veins; R – radial vein; Sc – subcosta; St – stigma). Scale bars = 0.5 mm.

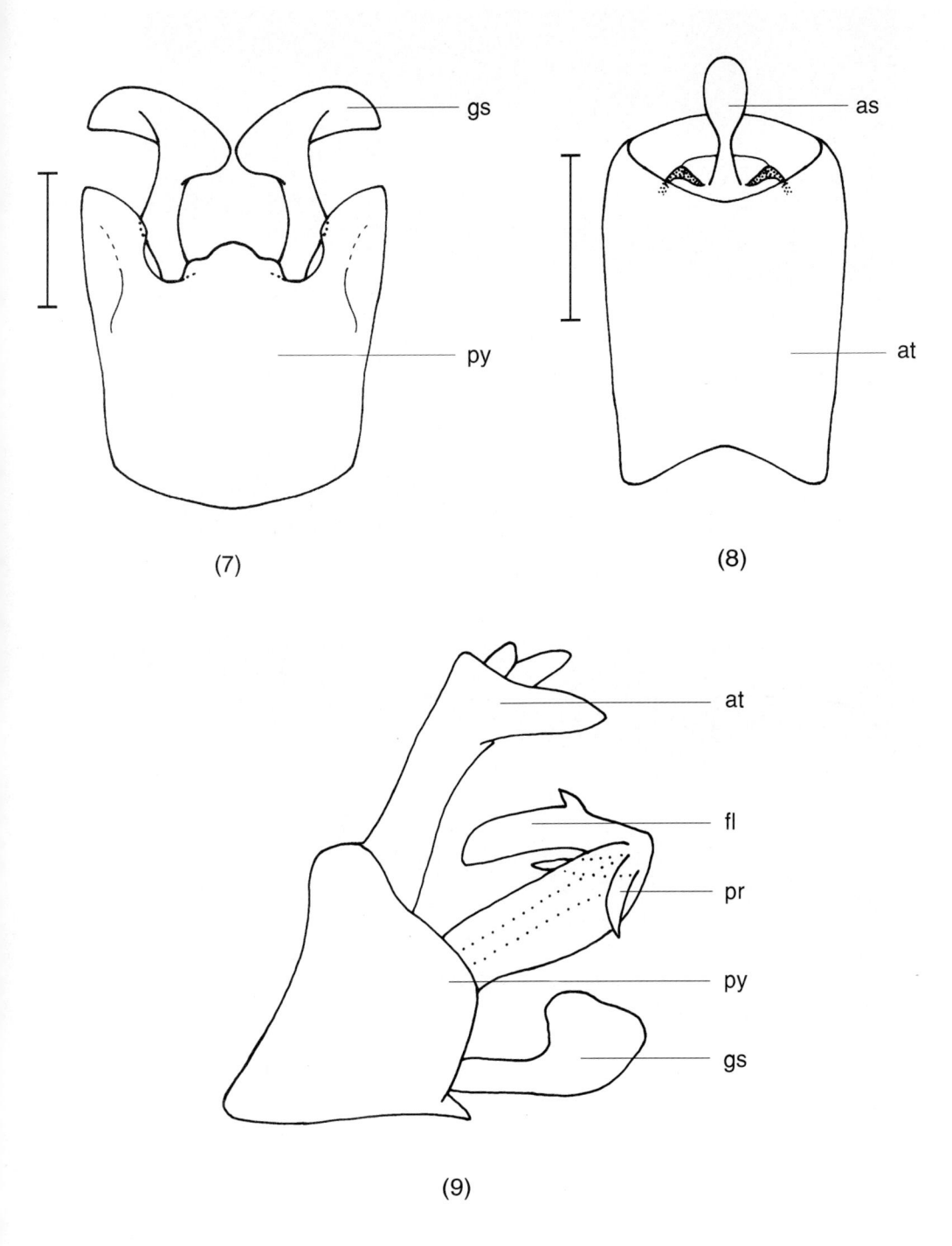

**Fig. 7–9** Diagnostic features of the external morphology, *Semo clypeatus*: (7) male pygofer, ventral view (gs – genital style; py – pygofer); (8) male anal tube, dorsal view (as – anal style; at – anal tube); (9) schematic representation of male genitalia (modified from Kramer (1981)), lateral view (at – anal tube; fl – flagellum; gs – genital style; pr – process of aedeagus; py – pygofer). Scale bars = 0.25 mm.

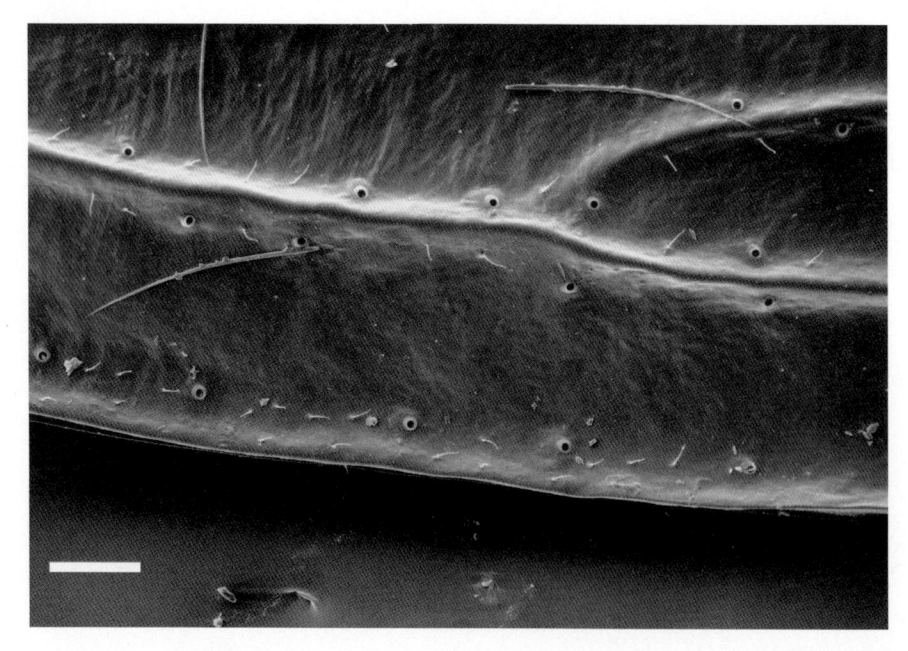

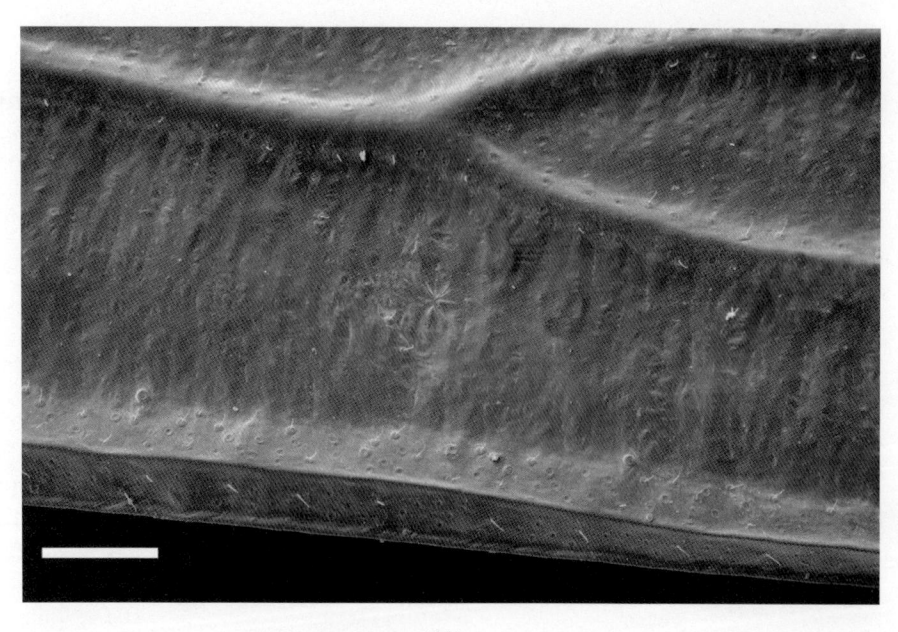

**Fig. 10–11** Scanning electron micrographs showing part of the forewings: (10) *Aka finitima*, veins with setiferous peduncles; (11) *Semo clypeatus*, veins without setiferous peduncles. Scale bar = 0.1 mm.

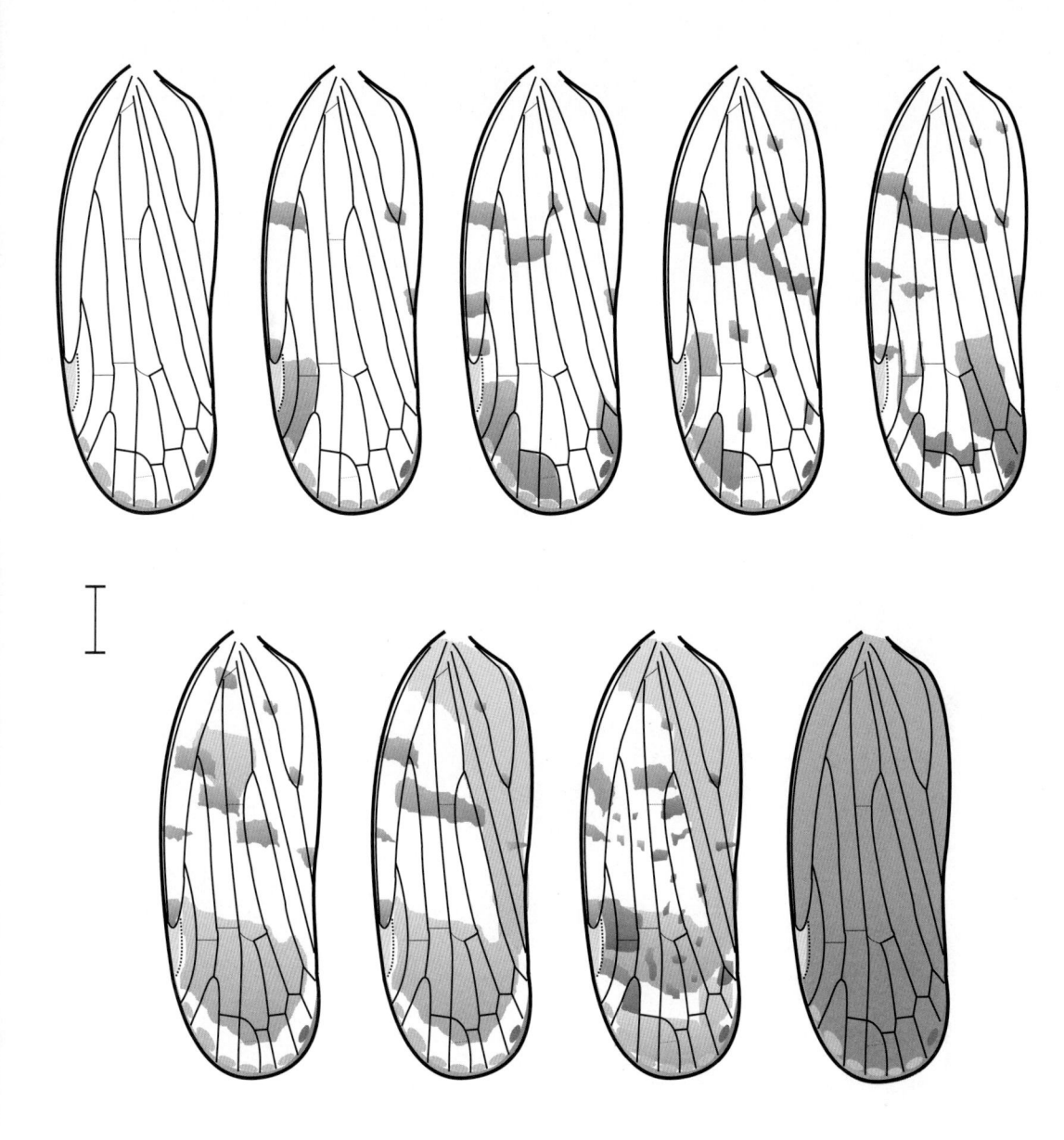

Fig. 12 *Semo*, forewing colour pattern variation. Scale bar = 0.5 mm.

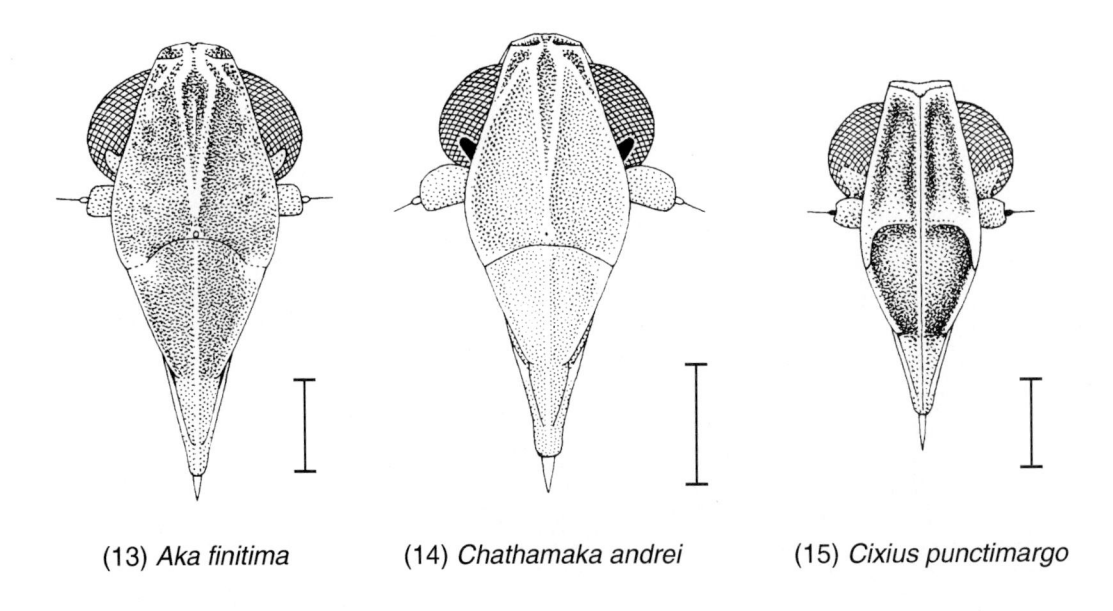

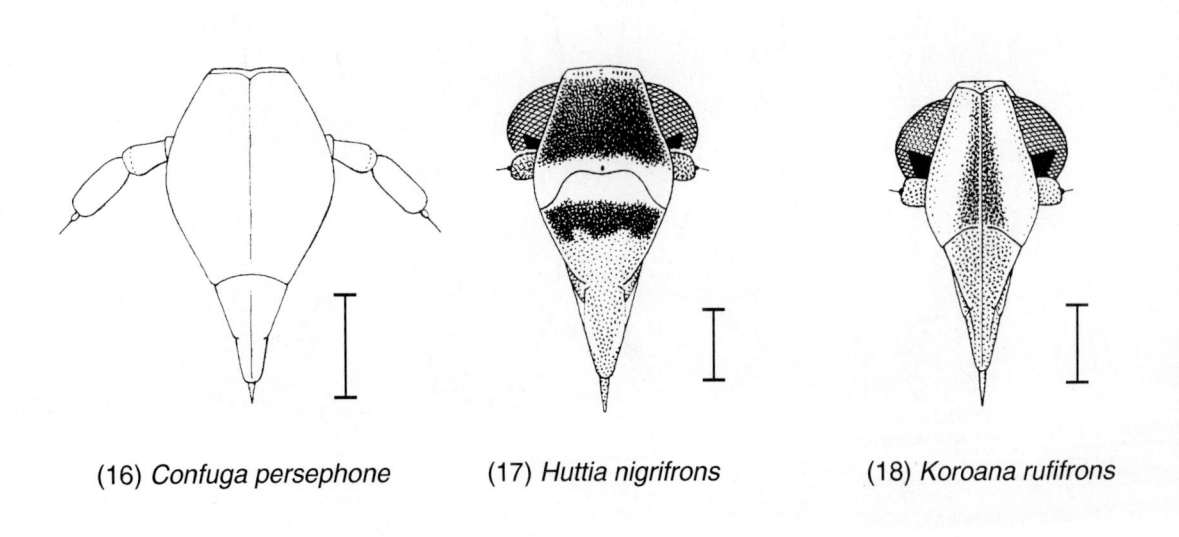

(13) *Aka finitima*     (14) *Chathamaka andrei*     (15) *Cixius punctimargo*

(16) *Confuga persephone*     (17) *Huttia nigrifrons*     (18) *Koroana rufifrons*

**Fig. 13–18** Head, frontal view. Scale bars = 0.5 mm.

*Fauna of New Zealand 40*

(19) *Malpha cockcrofti*

(20) *M. muiri*

(21) *Parasemo hutchesoni*

(22) *Semo clypeatus*

(23) *Tiriteana clarkei*

(24) *Oliarus atkinsoni*

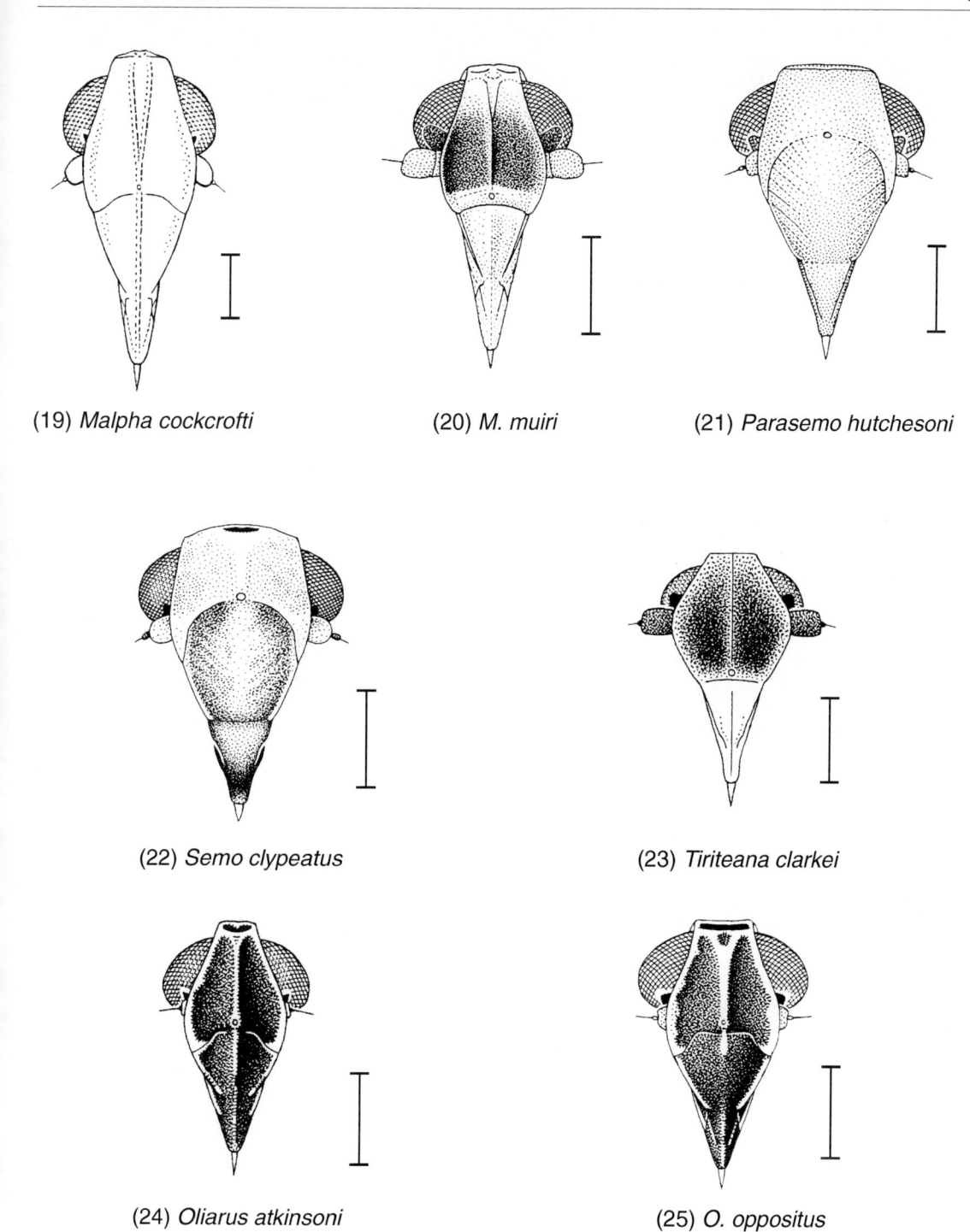

(25) *O. oppositus*

**Fig. 19–25** Head, frontal view. Scale bars = 0.5 mm.

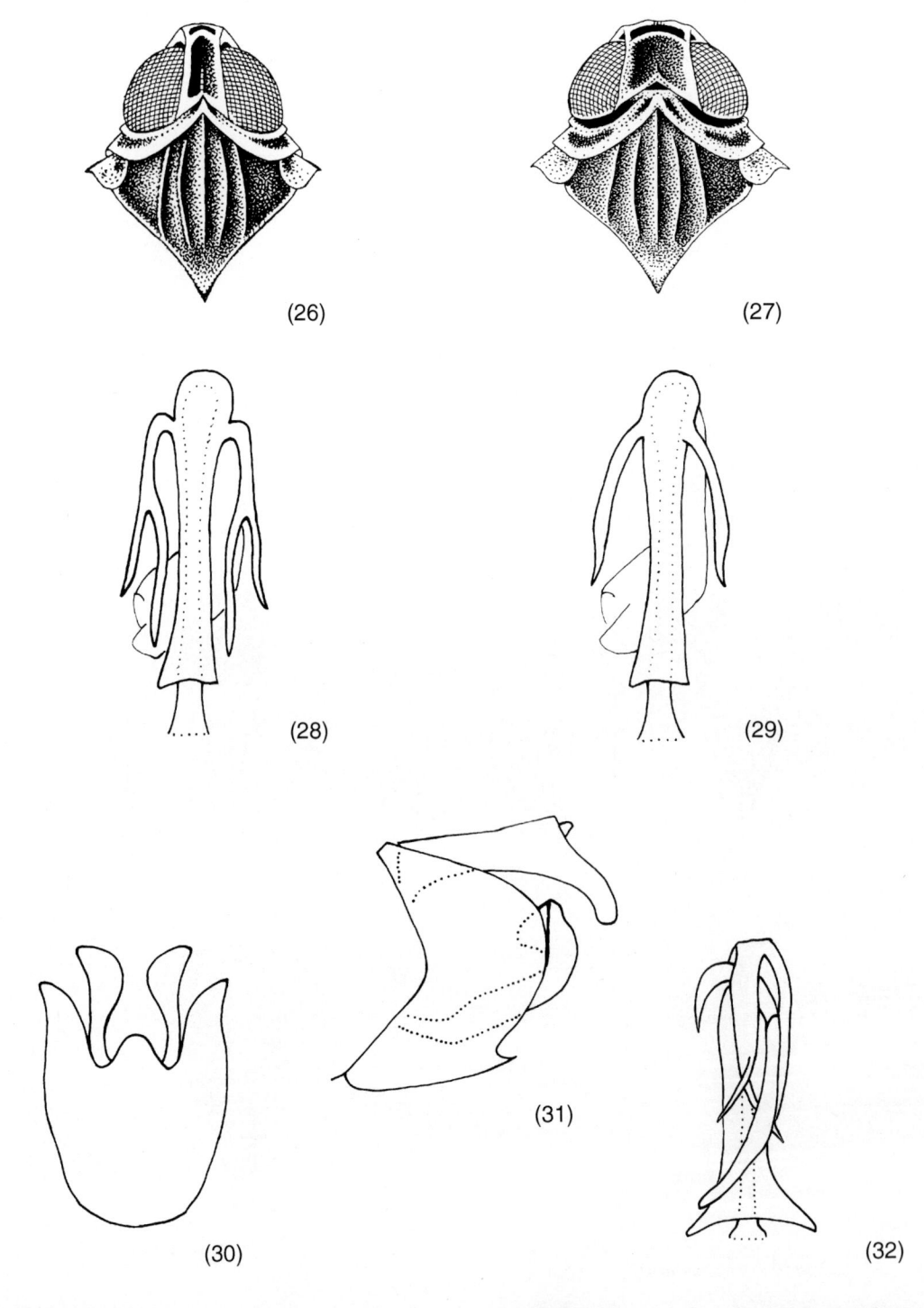

**Fig. 26–27** Head and thorax, dorsal view. (26) *Oliarus atkinsoni*; (27) *O. oppositus*. Scale bars = 1.0 mm. **Fig. 28–29** Sketch of aedeagus, ventral view: (28) hypothesised *Cixius* hybrid with 2 forked spinose processes; (29) hypothesised *Cixius* hybrid with 2 simple spinose processes. **Fig. 30–32** *Confuga persephone*, male genitalia (after Fennah 1975): (30) pygofer, ventral view; (31) pygofer, genital style, and anal tube, lateral view; (32) aedeagus, dorsal view.

Fauna of New Zealand 40

(33) *Aka finitima*

(34) *A. dunedinensis*

(35) *A. westlandica*

(36) *A. duniana*

(37) *A. rhodei*

(38) *Chathamaka andrei*

(39) *Cixius inexspectatus*

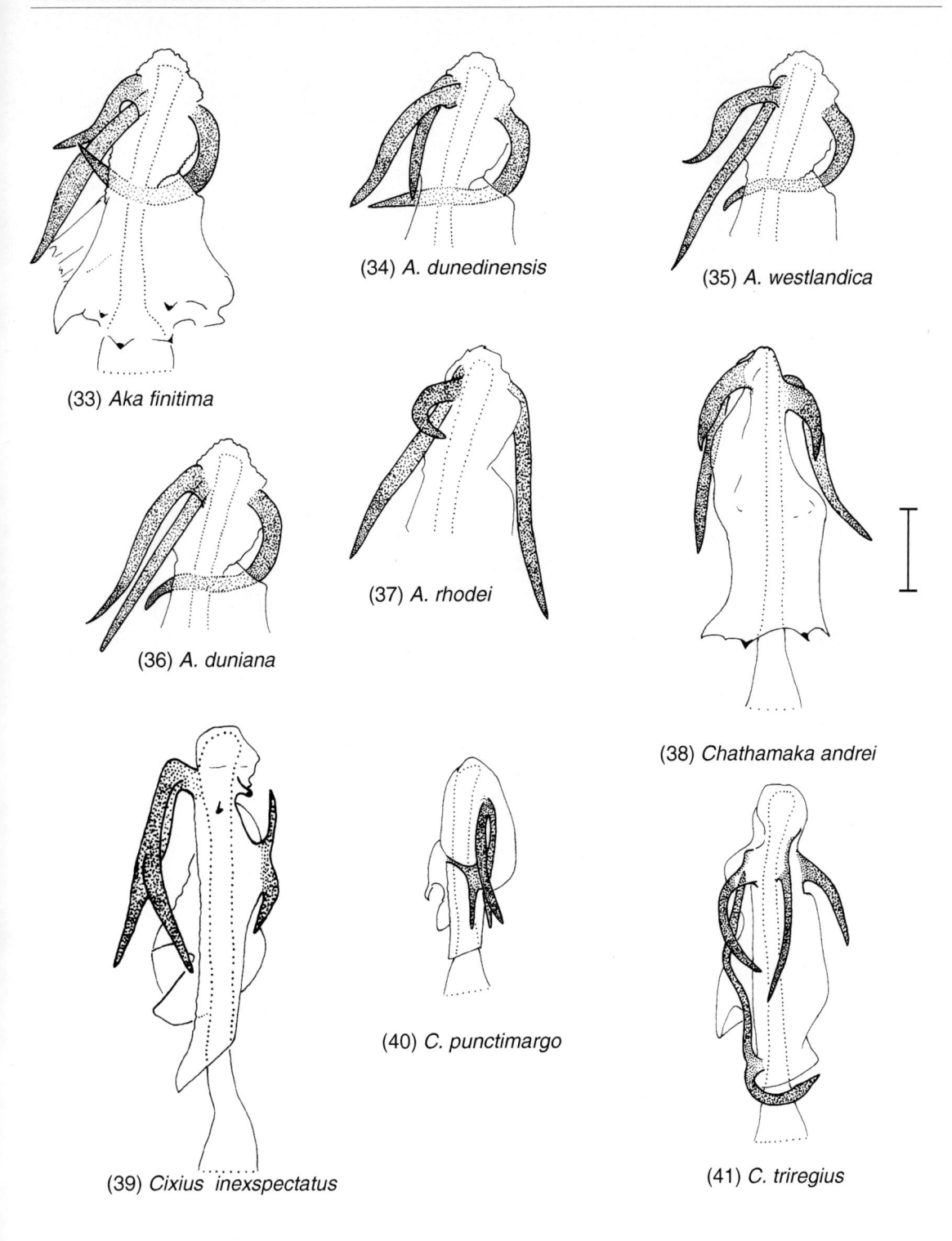

(40) *C. punctimargo*

(41) *C. triregius*

**Fig. 33–41** Aedeagus, ventral view. Scale bar = 0.1 mm.

(42) *Huttia nigrifrons*

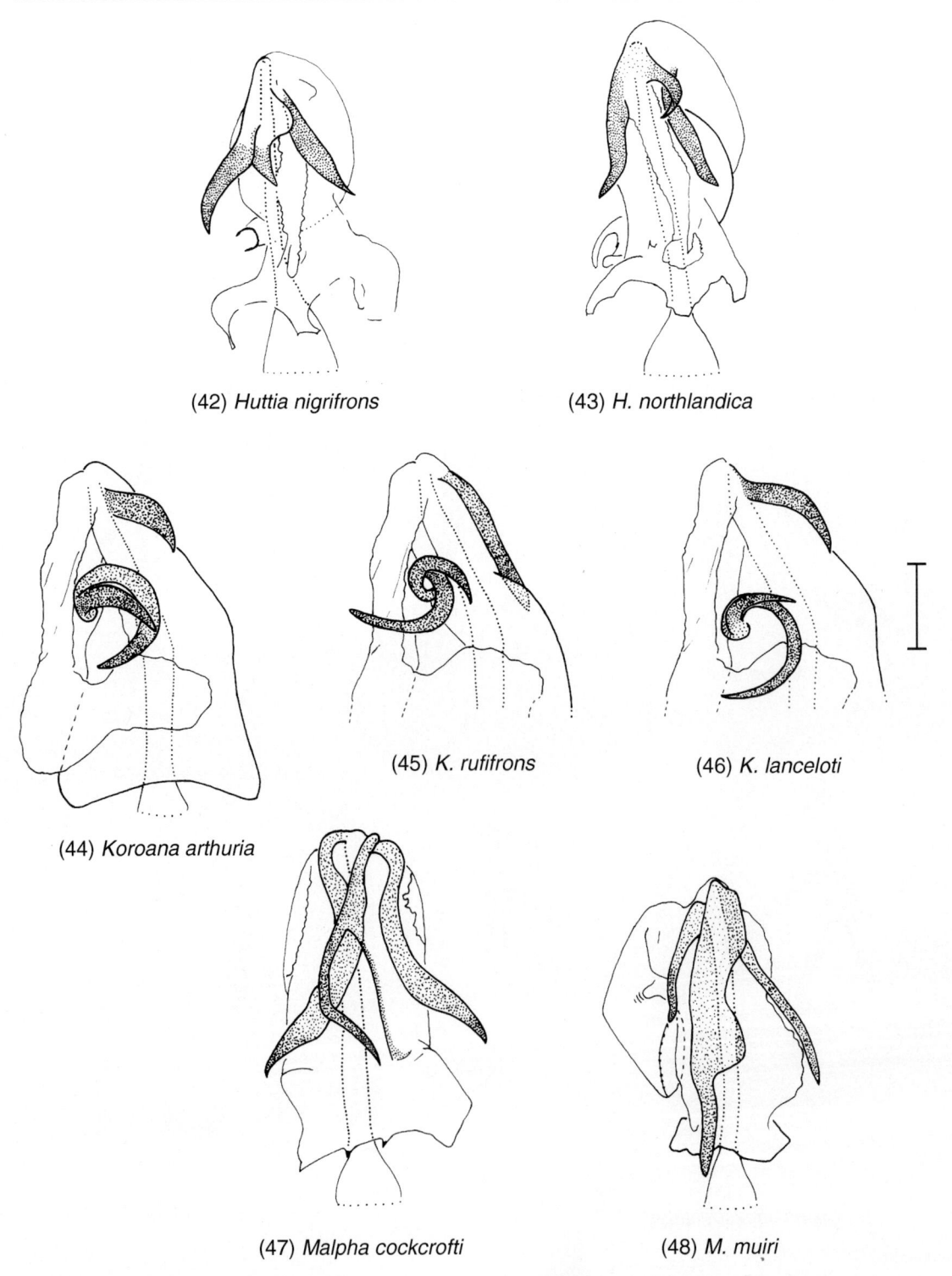

(43) *H. northlandica*

(44) *Koroana arthuria*

(45) *K. rufifrons*

(46) *K. lanceloti*

(47) *Malpha cockcrofti*

(48) *M. muiri*

**Fig. 42–43** Aedeagus, ventral view. Scale bar = 0.1 mm. **Fig. 44–46** Aedeagus, dorsal view. Scale bar = 0.1 mm. **Fig. 47–48** Aedeagus, ventral view. Scale bar = 0.1 mm.

Fauna of New Zealand 40

(49) *Parasemo hutchesoni*

(50) *Semo clypeatus*

(51) *S. southlandiae*

(52) *S. transinsularis*

(53) *S. harrisi*

(54) *Tiriteana clarkei*

(55) *Oliarus atkinsoni*

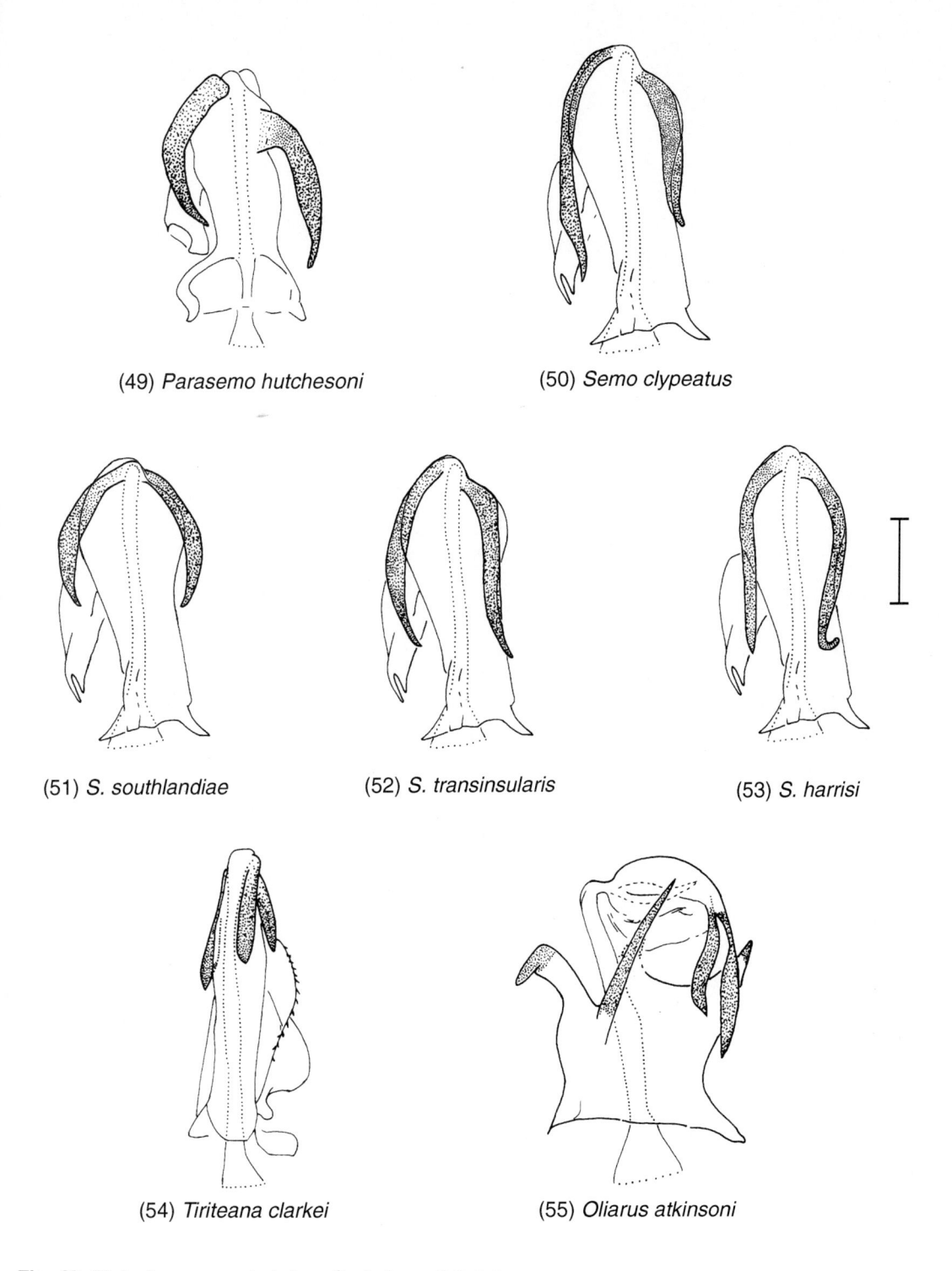

**Fig. 49–55** Aedeagus, ventral view. Scale bar = 0.1 mm.

(56) *Aka finitima* (57) *A. dunedinensis* (58) *A. westlandica* (59) *A. duniana*

(60) *A. rhodei* (61) *Chathamaka andrei* (62) *Cixius inexspectatus* (63) *C. punctimargo*

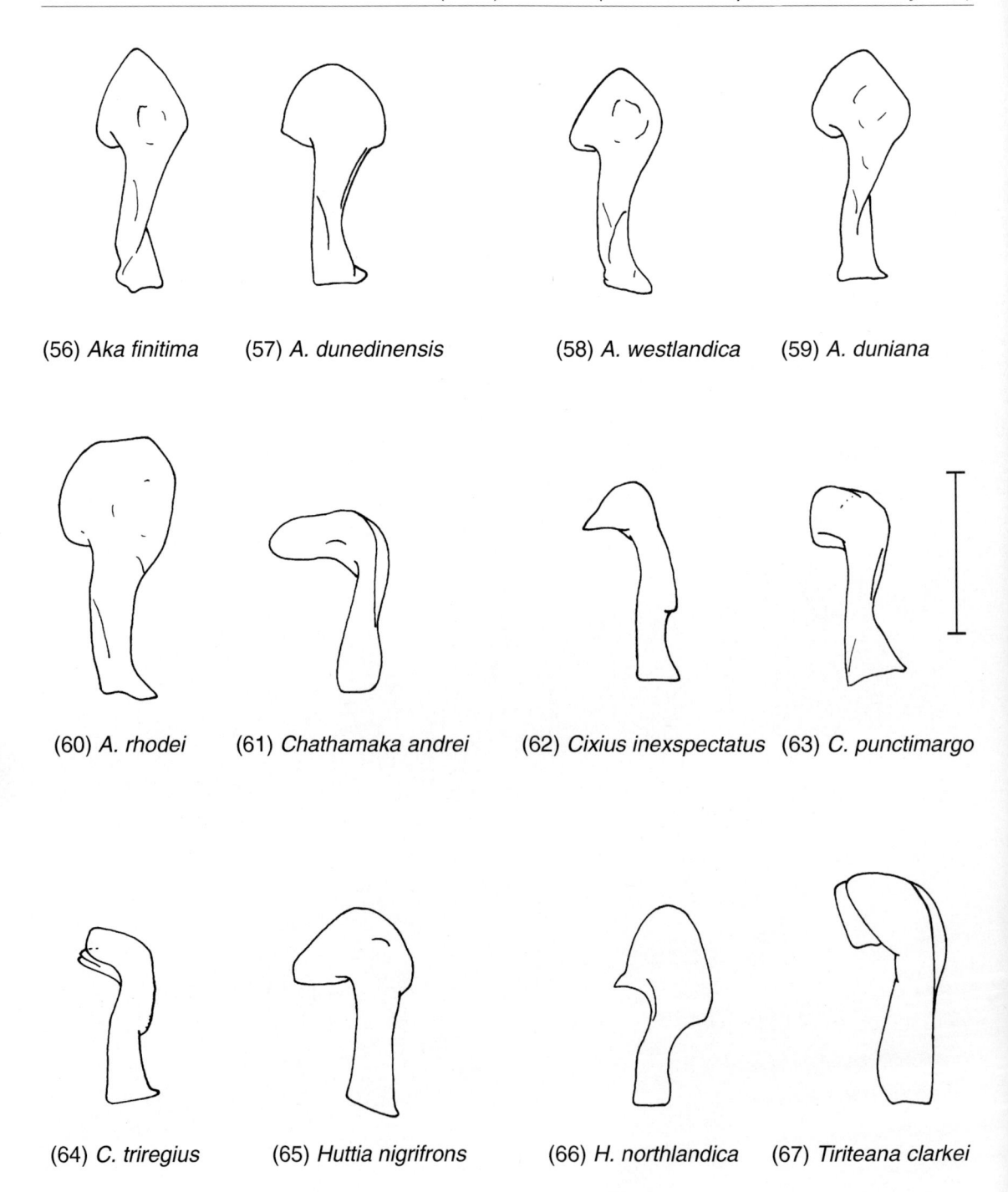

(64) *C. triregius* (65) *Huttia nigrifrons* (66) *H. northlandica* (67) *Tiriteana clarkei*

**Fig. 56–67** Left genital style, ventrad. Scale bar = 0.25 mm.

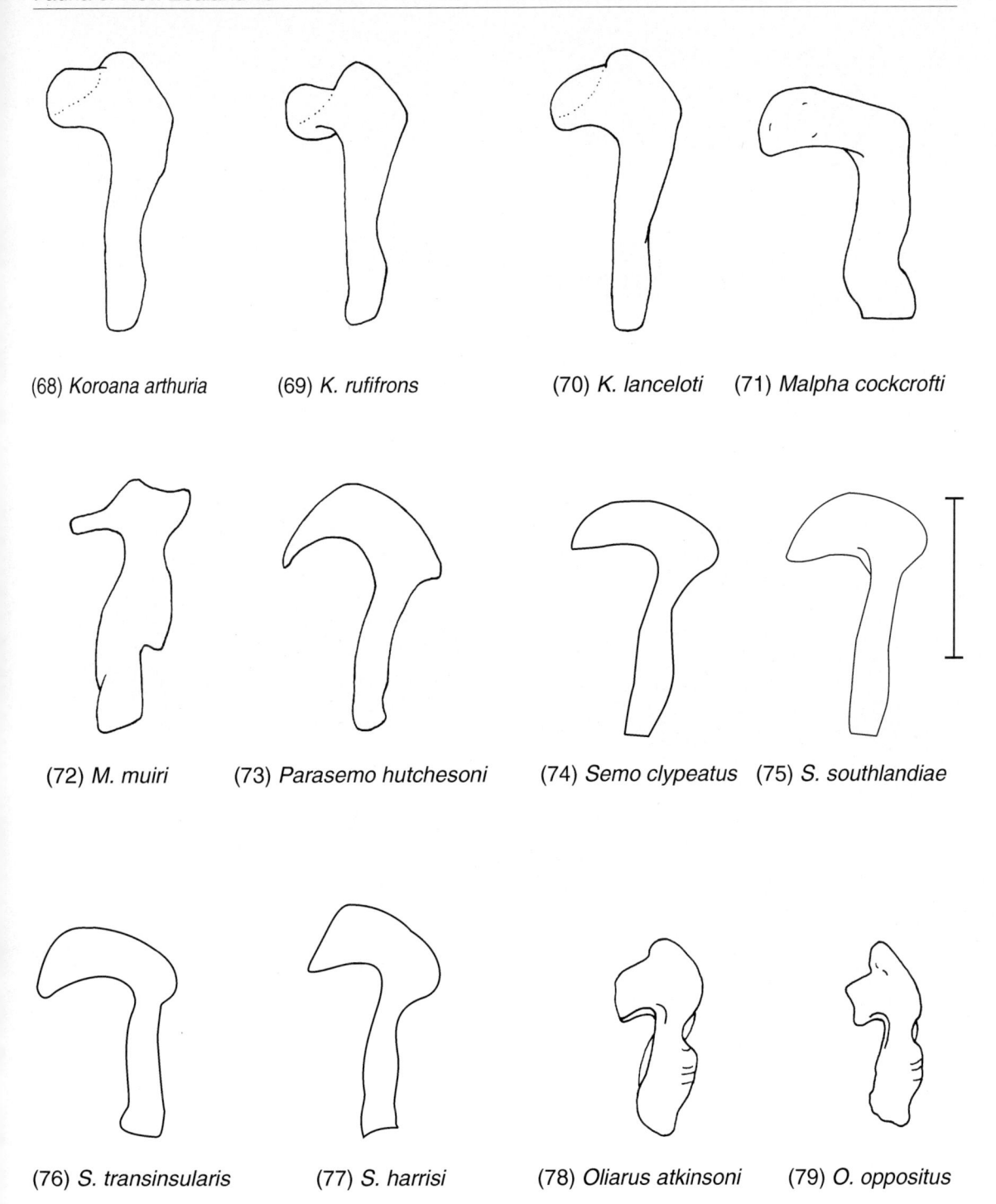

(68) *Koroana arthuria*   (69) *K. rufifrons*   (70) *K. lanceloti*   (71) *Malpha cockcrofti*

(72) *M. muiri*   (73) *Parasemo hutchesoni*   (74) *Semo clypeatus*   (75) *S. southlandiae*

(76) *S. transinsularis*   (77) *S. harrisi*   (78) *Oliarus atkinsoni*   (79) *O. oppositus*

**Fig. 68–79** Left genital style, ventrad. Scale bar = 0.25 mm.

(80) *Aka finitima*    (81) *Chathamaka andrei*    (82) *Cixius inexspectatus*    (83) *C. punctimargo*

(84) *Huttia nigrifrons*    (85) *Koroana arthuria*    (86) *Malpha muiri*

(87) *Parasemo hutchesoni*    (88) *Semo clypeatus*    (89) *Tiriteana clarkei*    (90) *Oliarus atkinsoni*

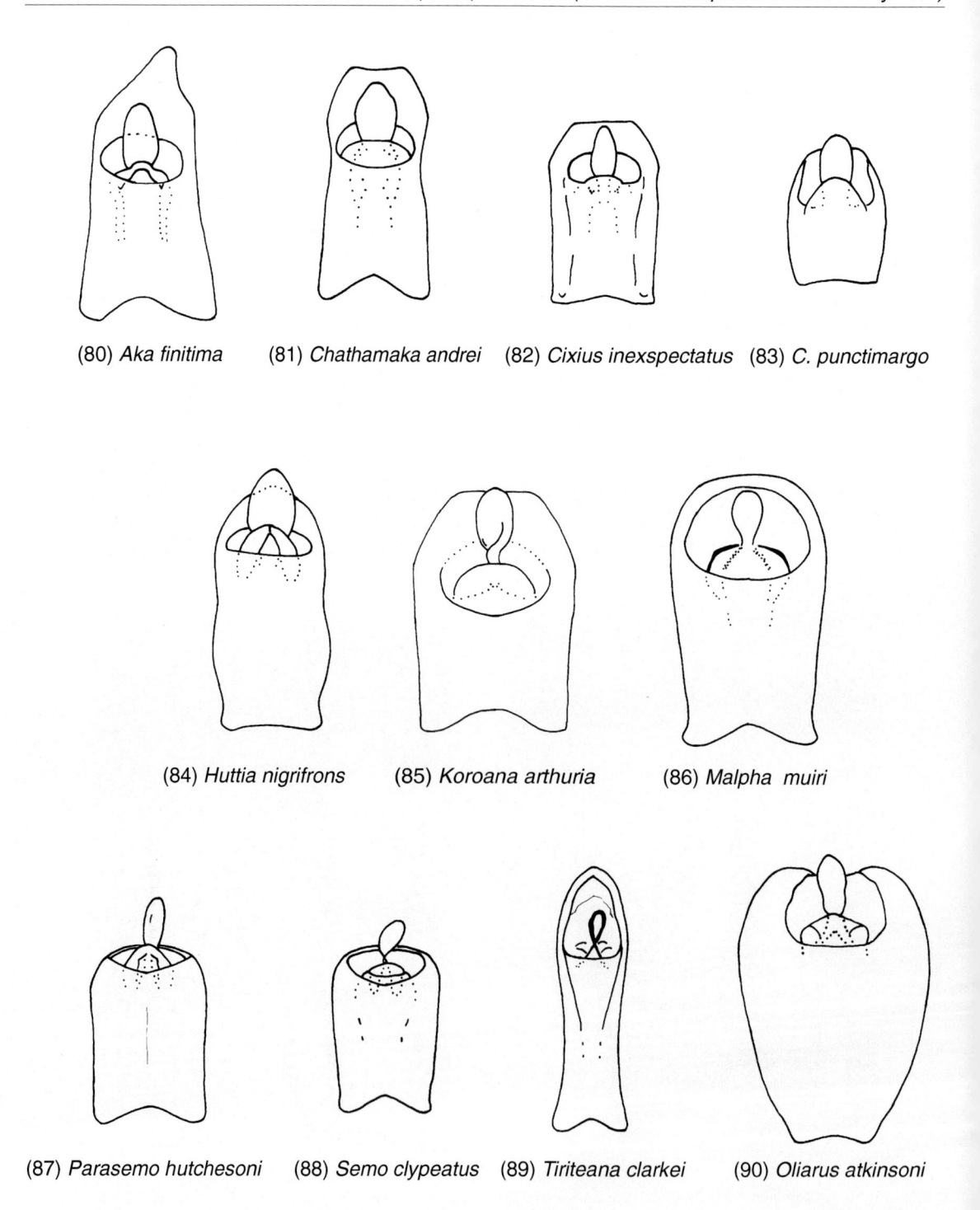

**Fig. 80–90** Male anal tube, dorsal view.

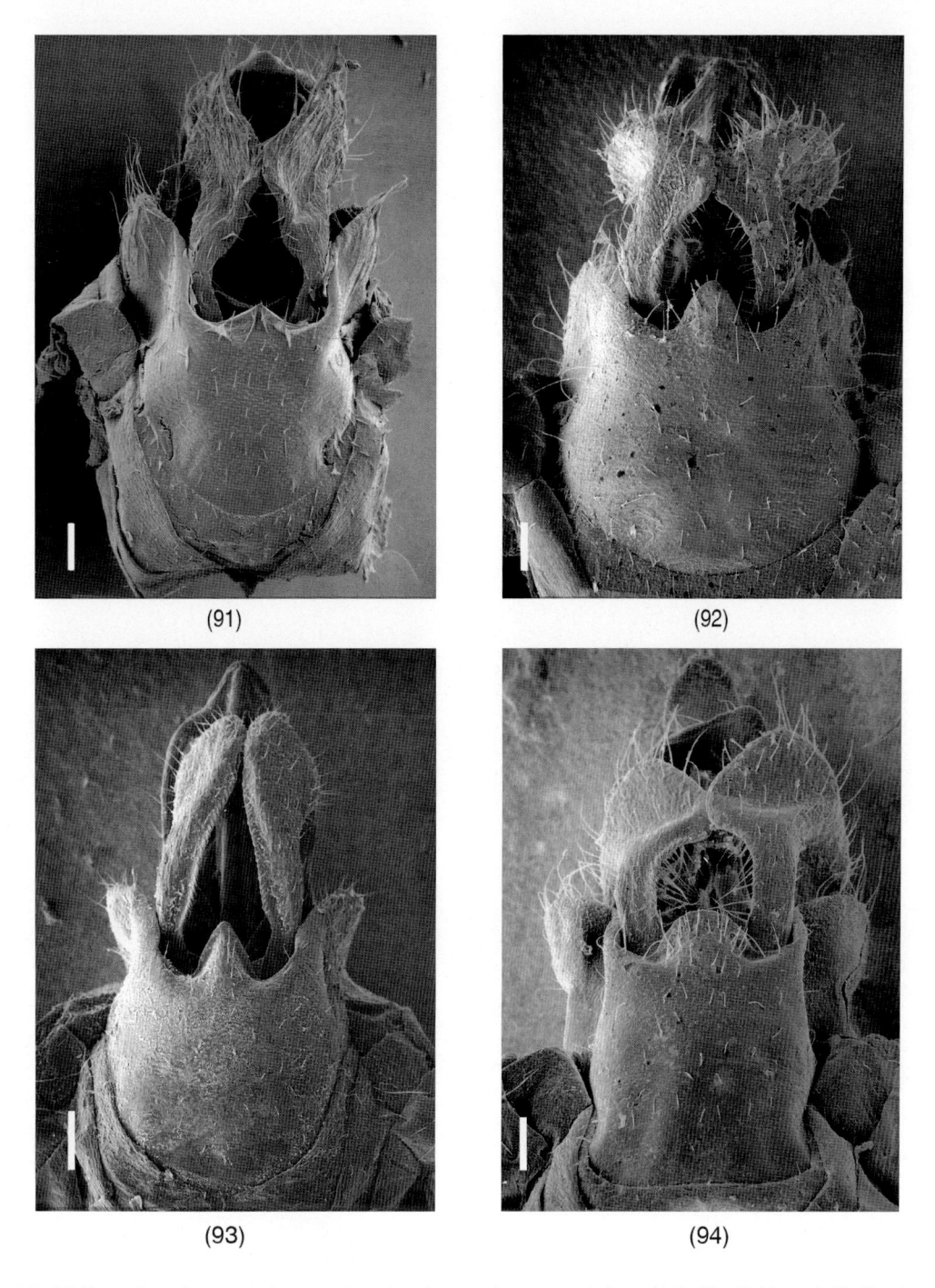

**Fig. 91–94** Scanning electron micrographs of male pygofer, ventral view: (91) *Aka finitima*; (92) *Chathamaka andrei*; (93) *Cixius inexspectatus*; (94) *Huttia nigrifrons*. Scale bars = 0.1 mm.

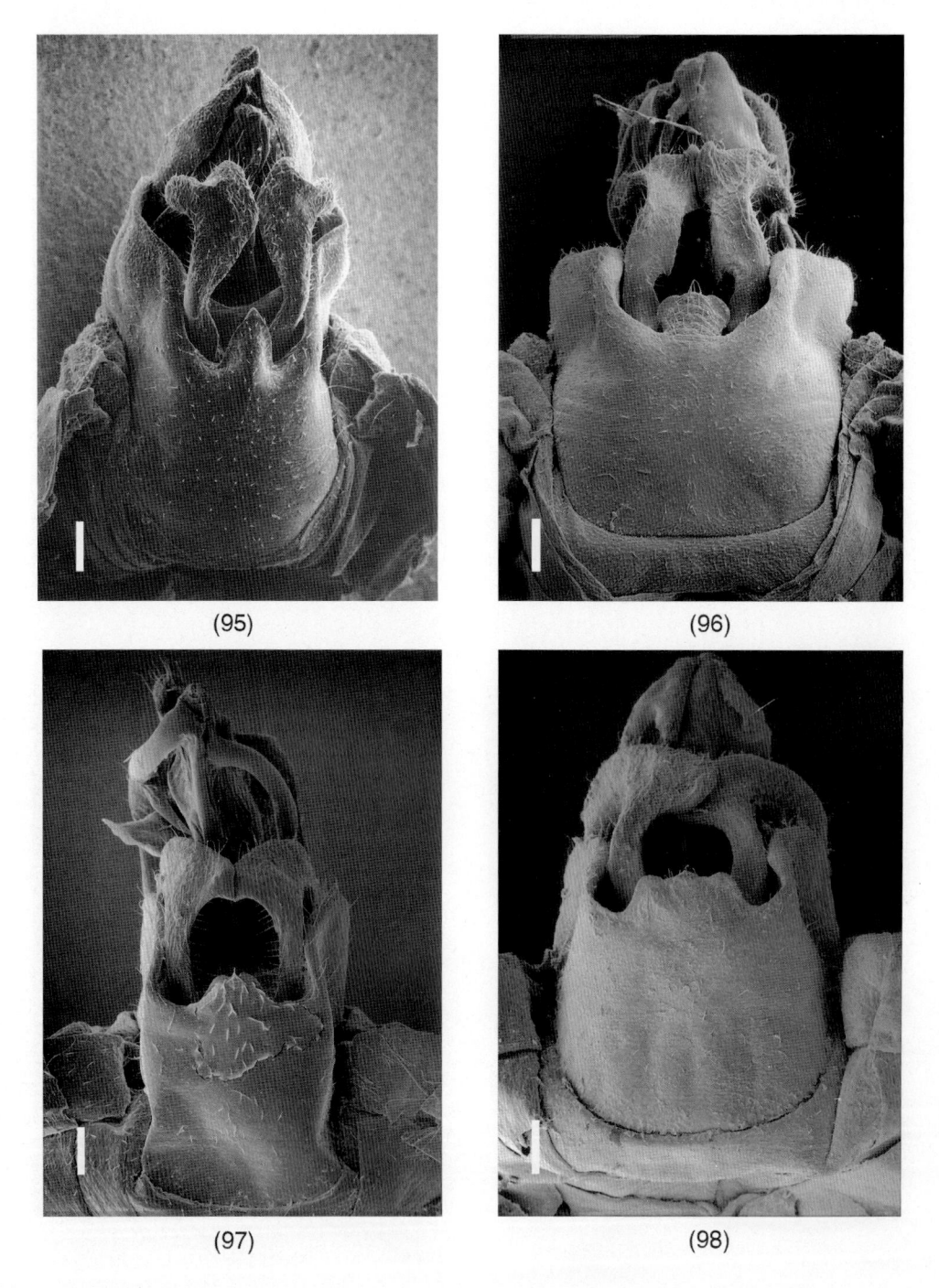

**Fig. 95–98** Scanning electron micrographs of male pygofer, ventral view: (95) *Koroana arthuria*; (96) *Malpha muiri*; (97) *Parasemo hutchesoni*; (98) *Semo clypeatus*. Scale bars = 0.1 mm.

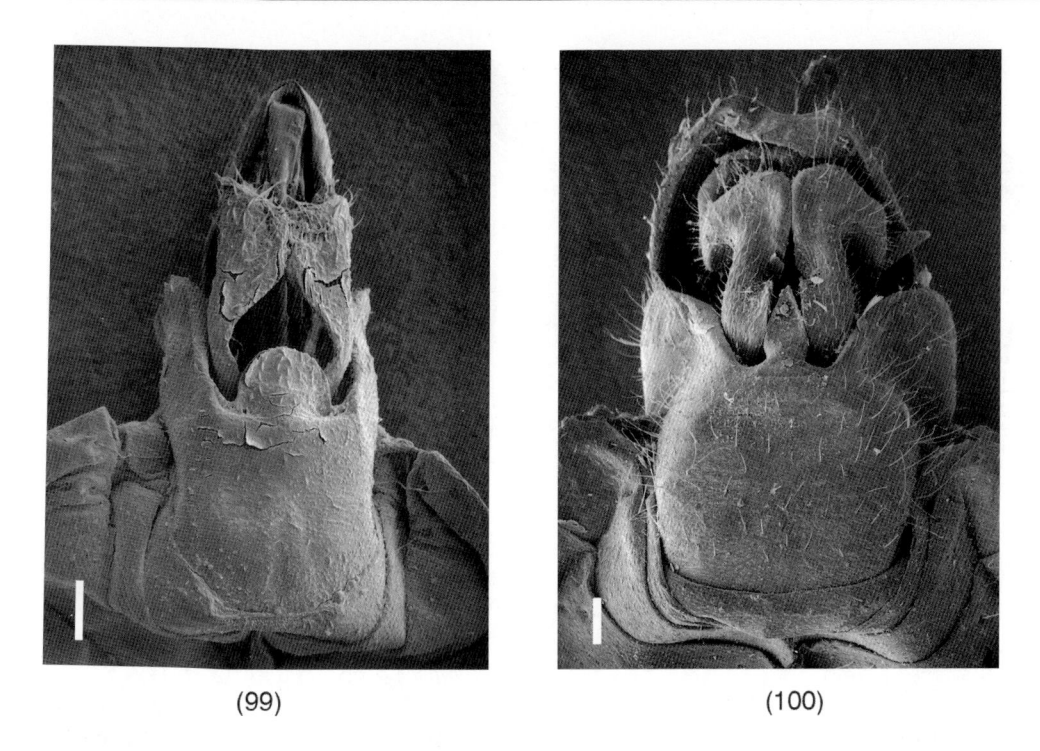

**Fig. 99–100** Scanning electron micrographs of male pygofer, ventral view: (99) *Tiriteana clarkei*; (100) *Oliarus atkinsoni*. Scale bars = 0.1 mm.

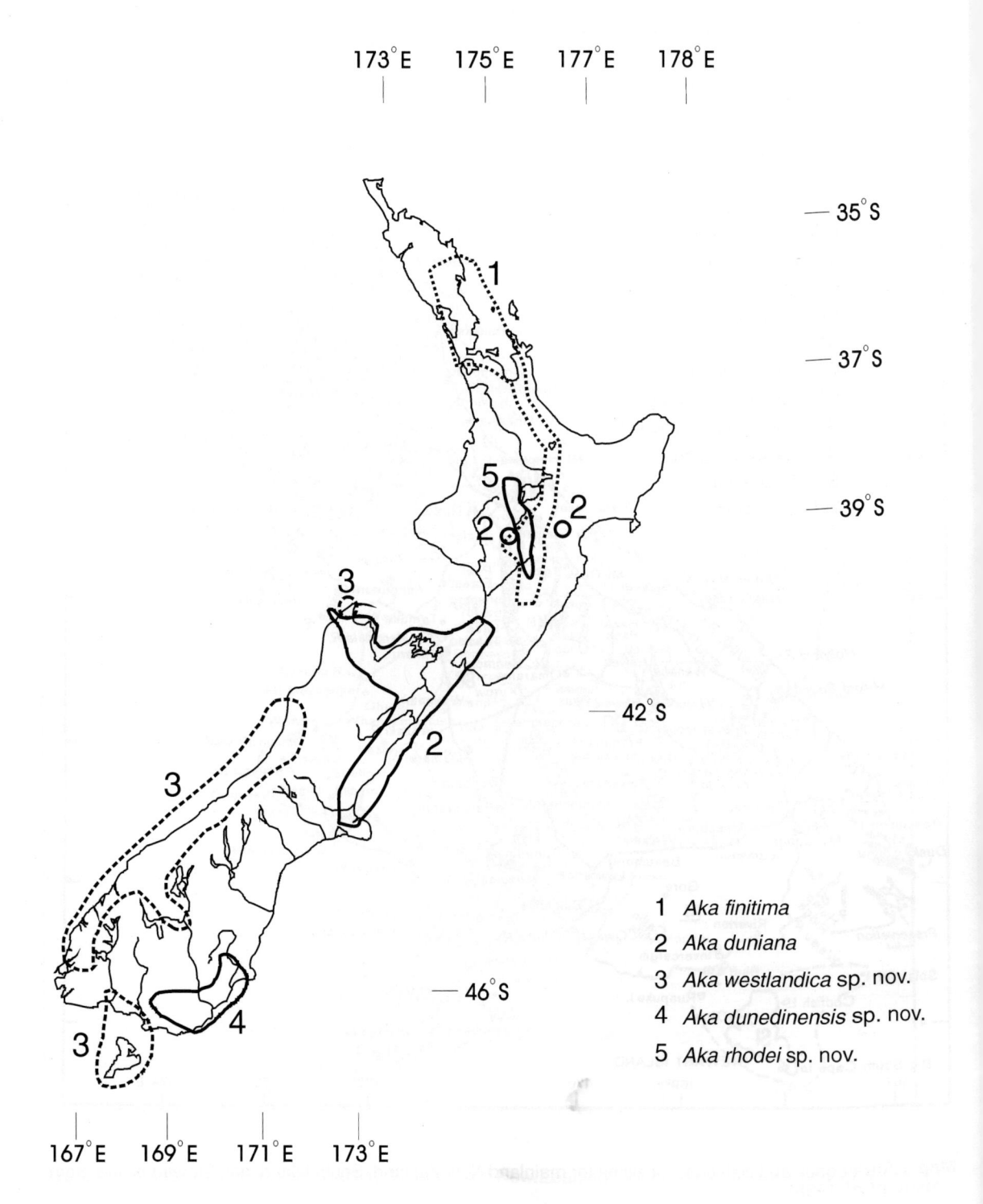

**Map 4** Distribution ranges of *Aka* species, mainland New Zealand.

1 *Aka finitima*
2 *Aka duniana*
3 *Aka westlandica* sp. nov.
4 *Aka dunedinensis* sp. nov.
5 *Aka rhodei* sp. nov.

Fauna of New Zealand 40

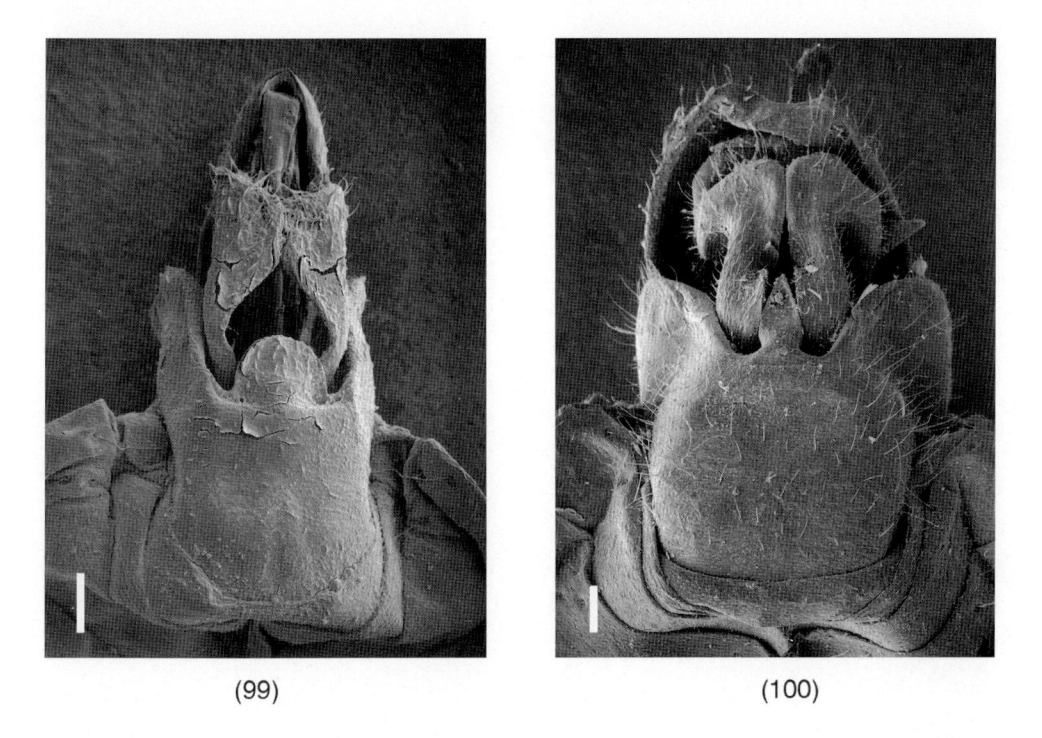

**Fig. 99–100** Scanning electron micrographs of male pygofer, ventral view: (99) *Tiriteana clarkei*; (100) *Oliarus atkinsoni*. Scale bars = 0.1 mm.

(101) *Aka finitima*

(102) *Chathamaka andrei*

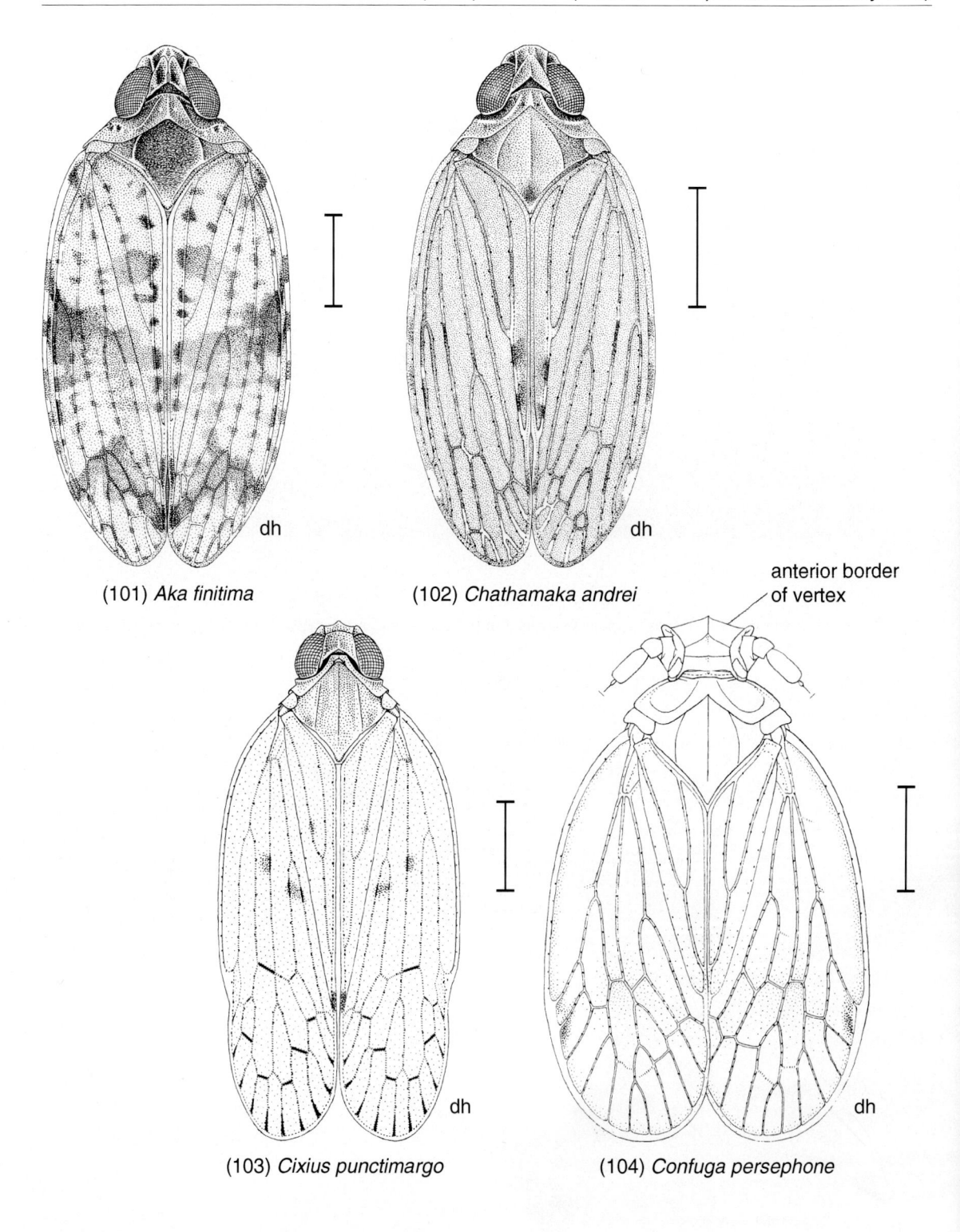

(103) *Cixius punctimargo*

(104) *Confuga persephone*

**Fig. 101–112** Habitus drawings of Cixiidae (Illustrator: D. W. Helmore). Scale bars = 1.0 mm.

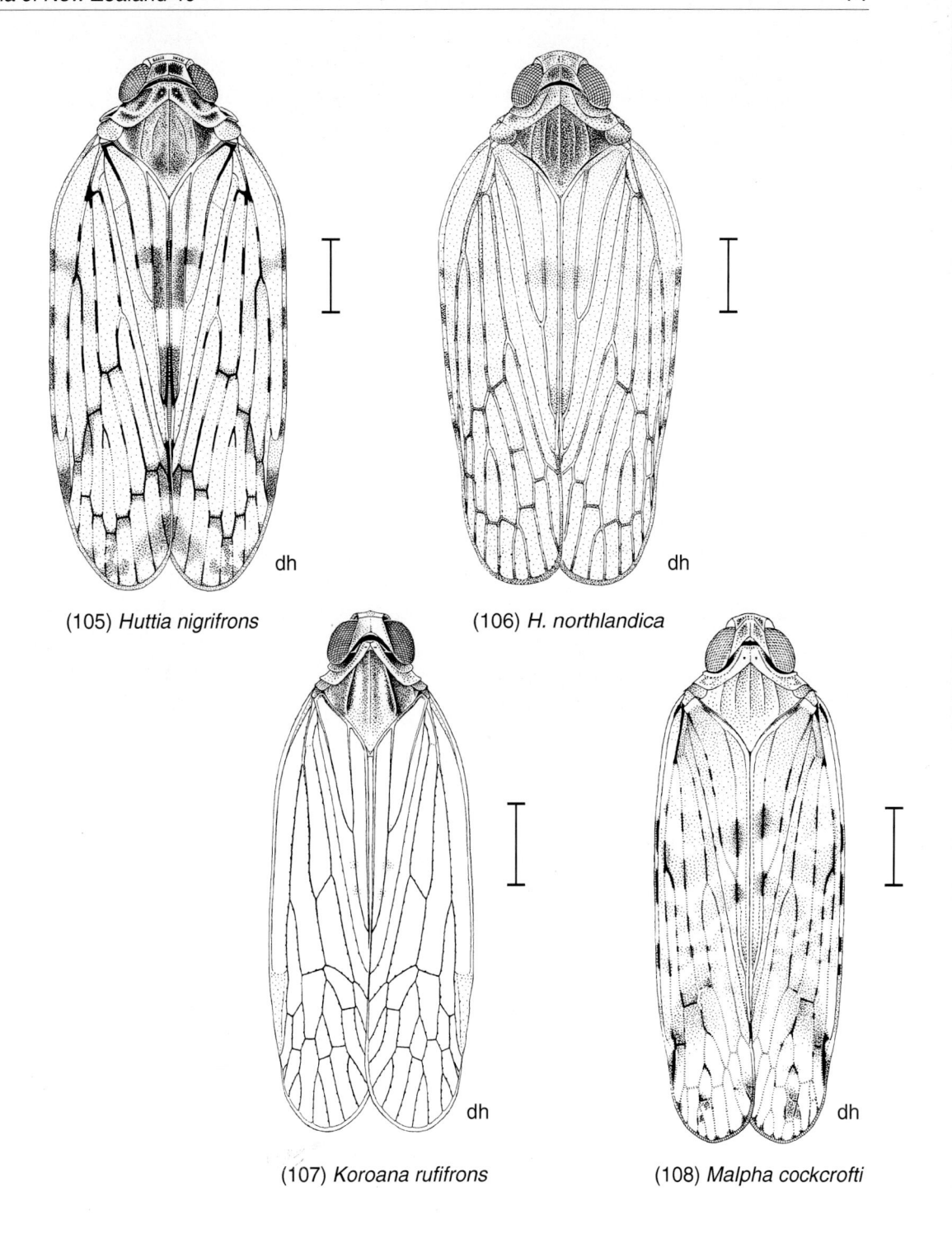

(105) *Huttia nigrifrons*

(106) *H. northlandica*

(107) *Koroana rufifrons*

(108) *Malpha cockcrofti*

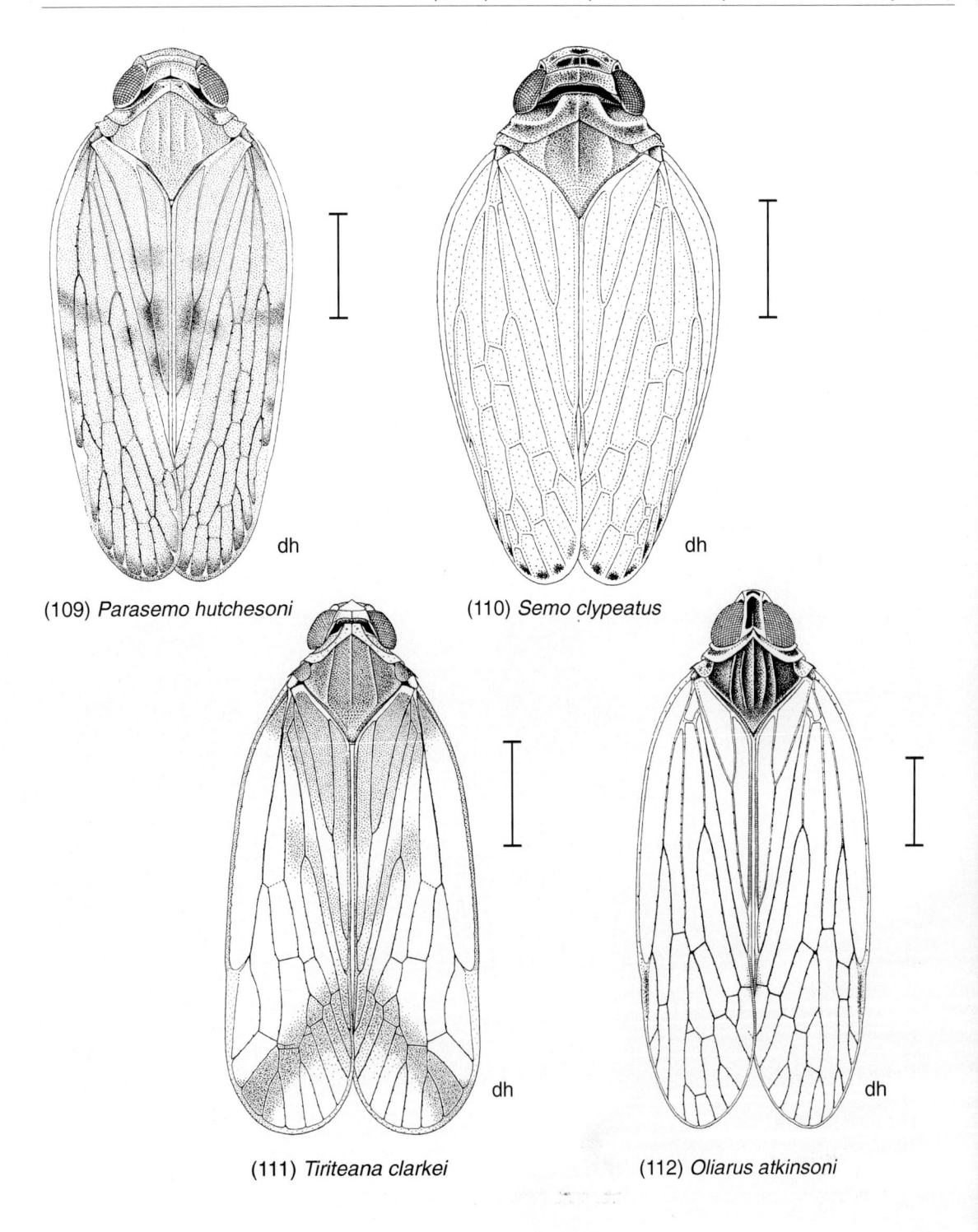

(109) *Parasemo hutchesoni*

(110) *Semo clypeatus*

(111) *Tiriteana clarkei*

(112) *Oliarus atkinsoni*

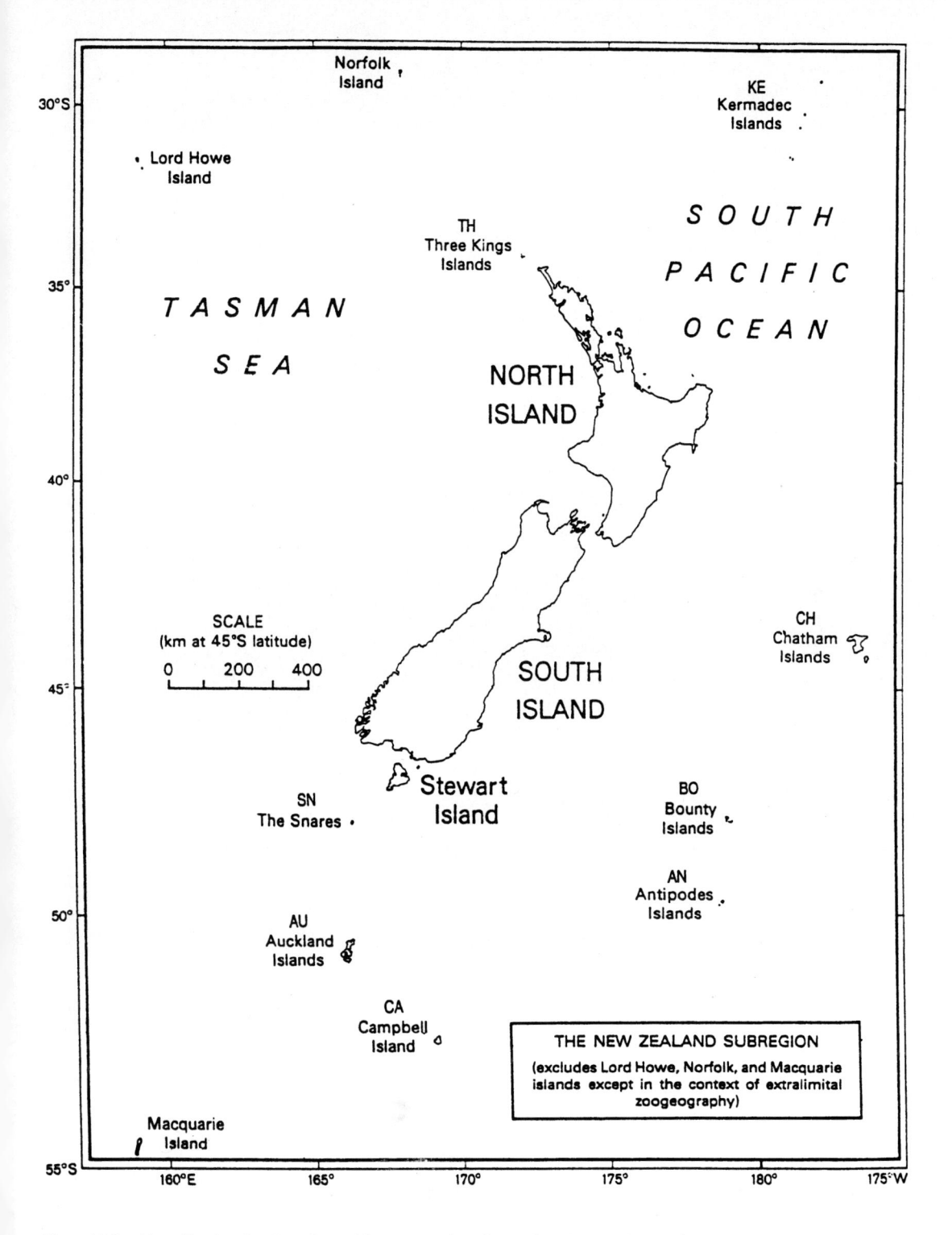

Map 1 The New Zealand subregion with area codes (from Crosby *et al.* 1998).

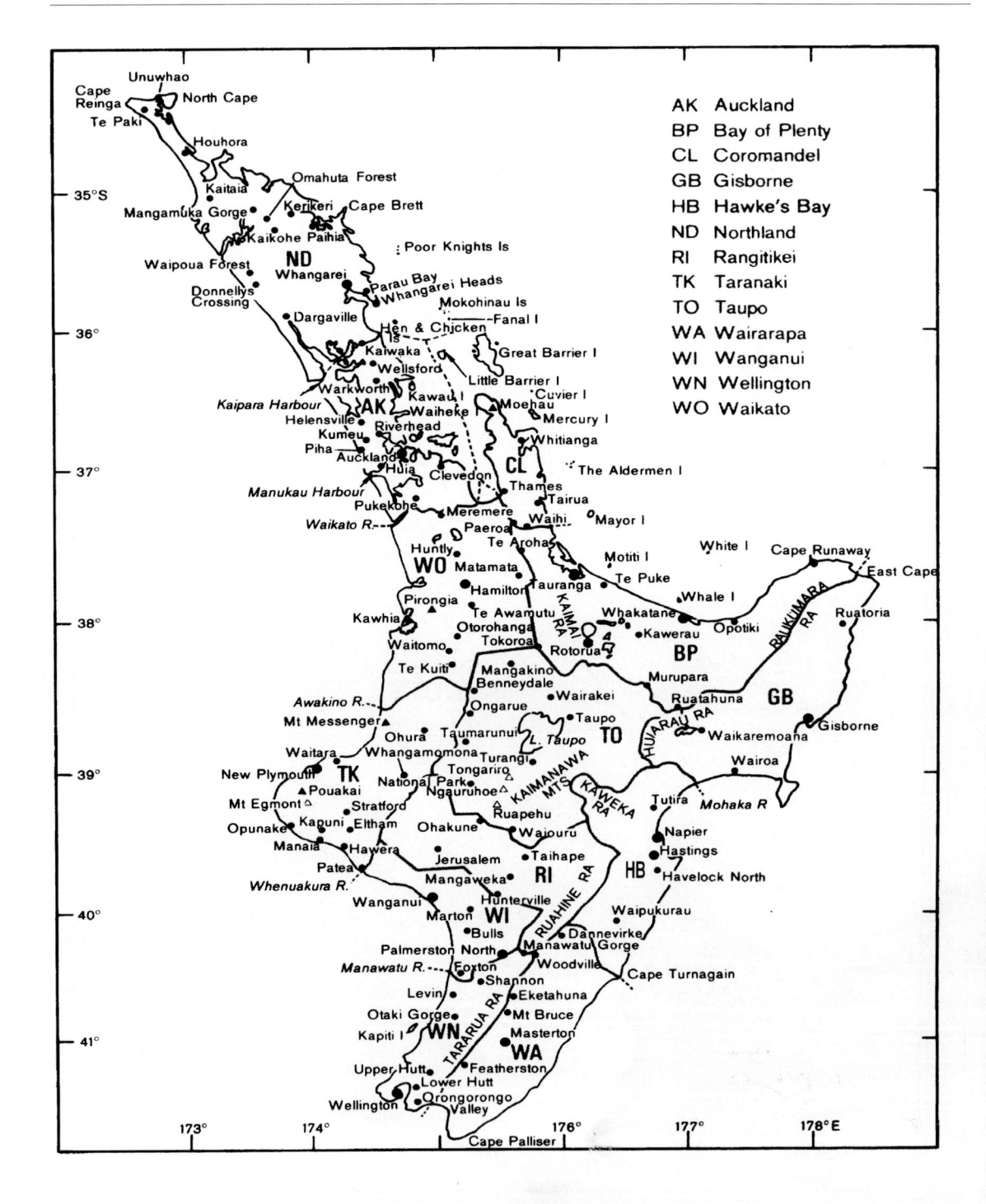

Map 2 Area codes and collecting localities for mainland New Zealand, North Island (from Crosby *et al.* 1998).

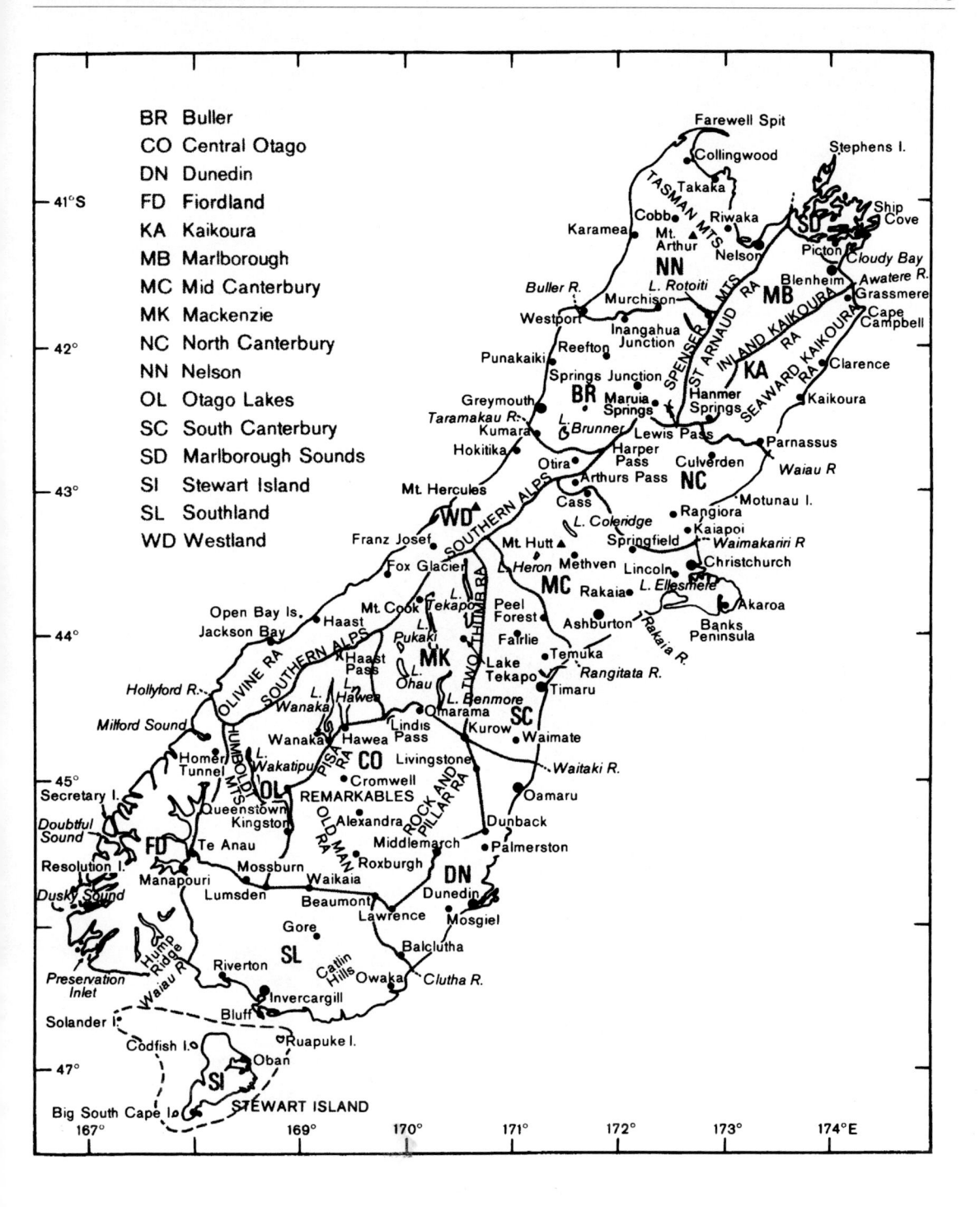

**Map 3** Area codes and collecting localities for mainland New Zealand, South Island and Stewart Island (from Crosby *et al.* 1998).

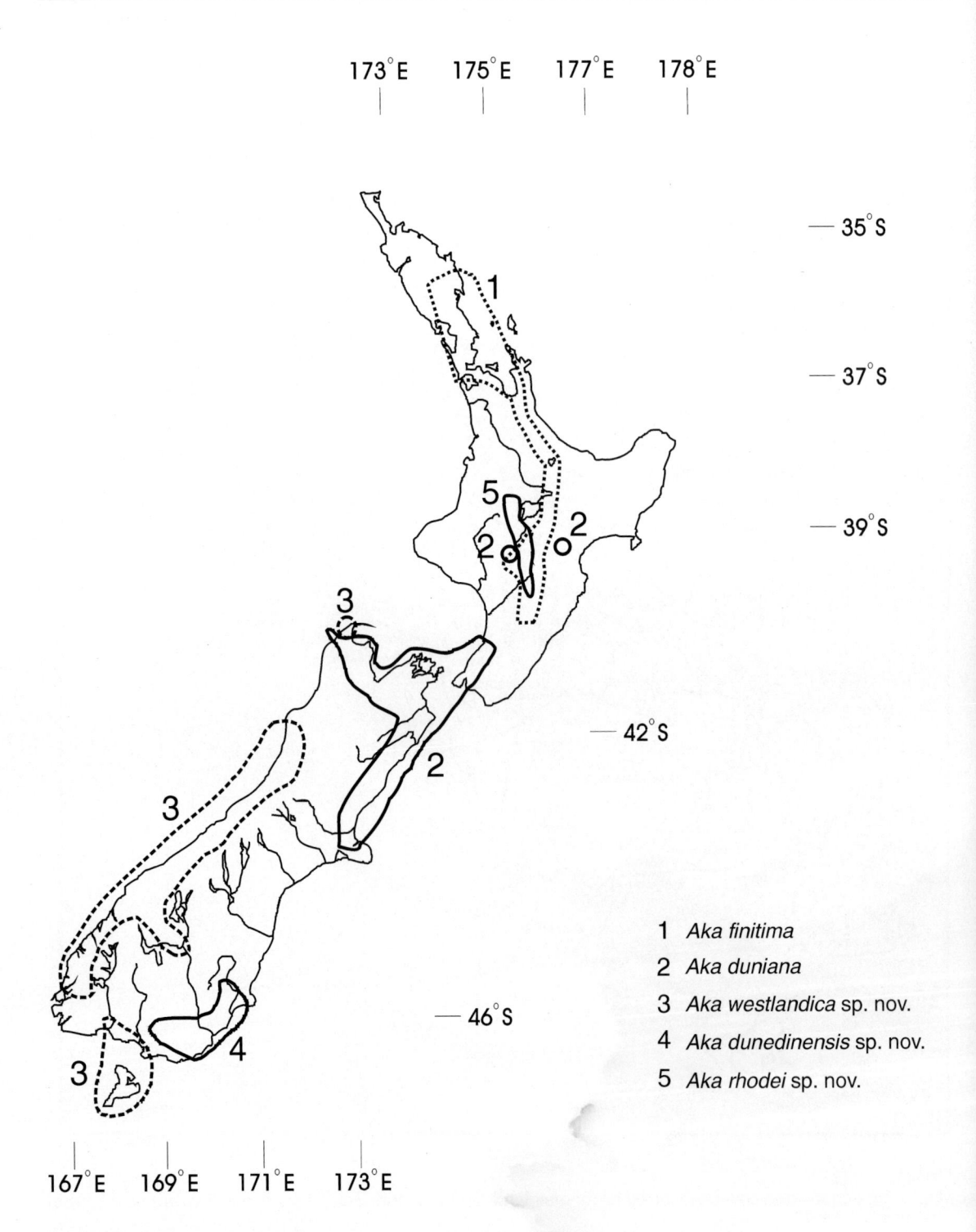

**Map 4** Distribution ranges of *Aka* species, mainland New Zealand.

1 *Aka finitima*
2 *Aka duniana*
3 *Aka westlandica* sp. nov.
4 *Aka dunedinensis* sp. nov.
5 *Aka rhodei* sp. nov.

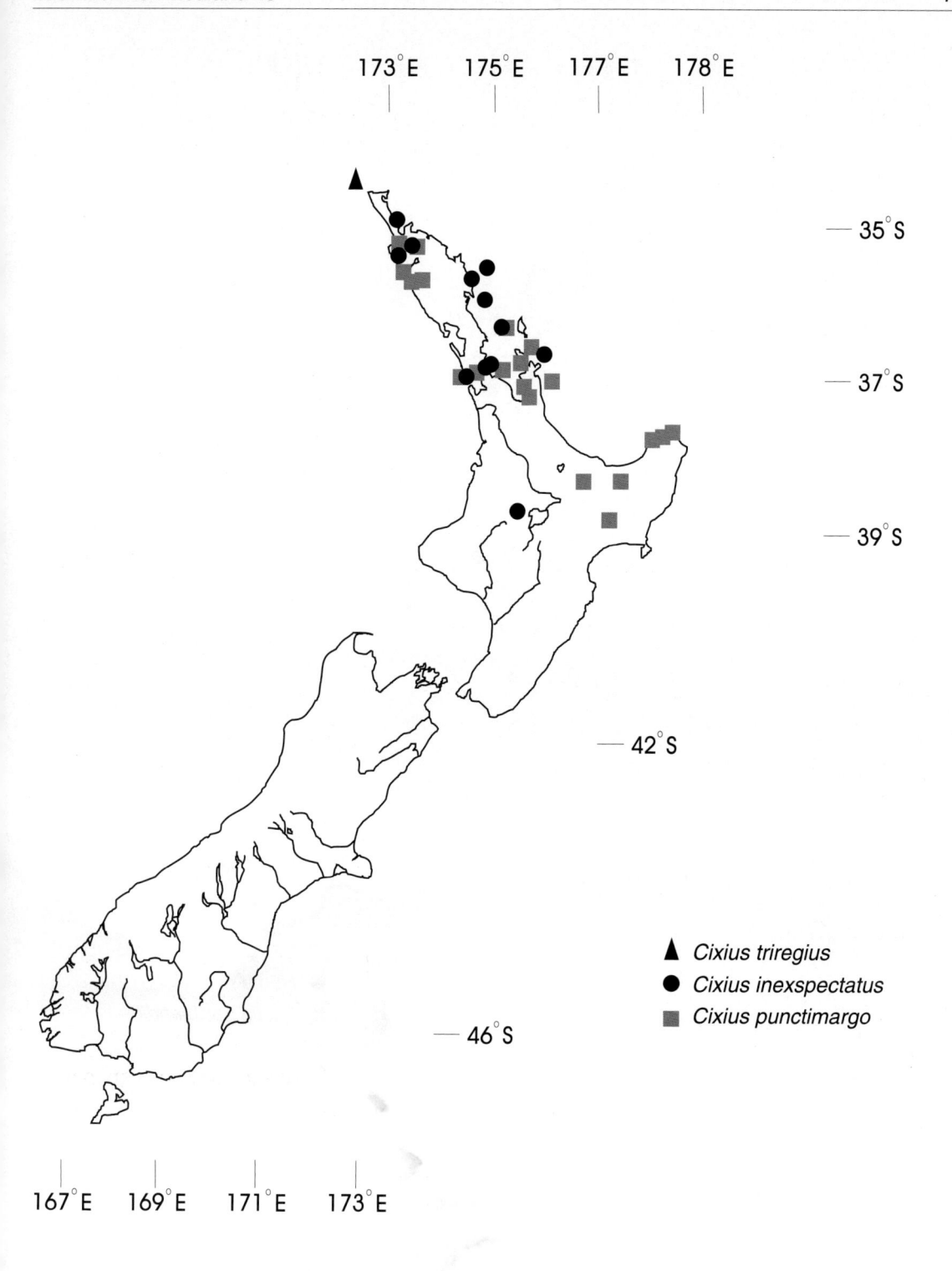

**Map 5** Collection localities of *Cixius* species, mainland New Zealand.

▲ *Cixius triregius*
● *Cixius inexspectatus*
■ *Cixius punctimargo*

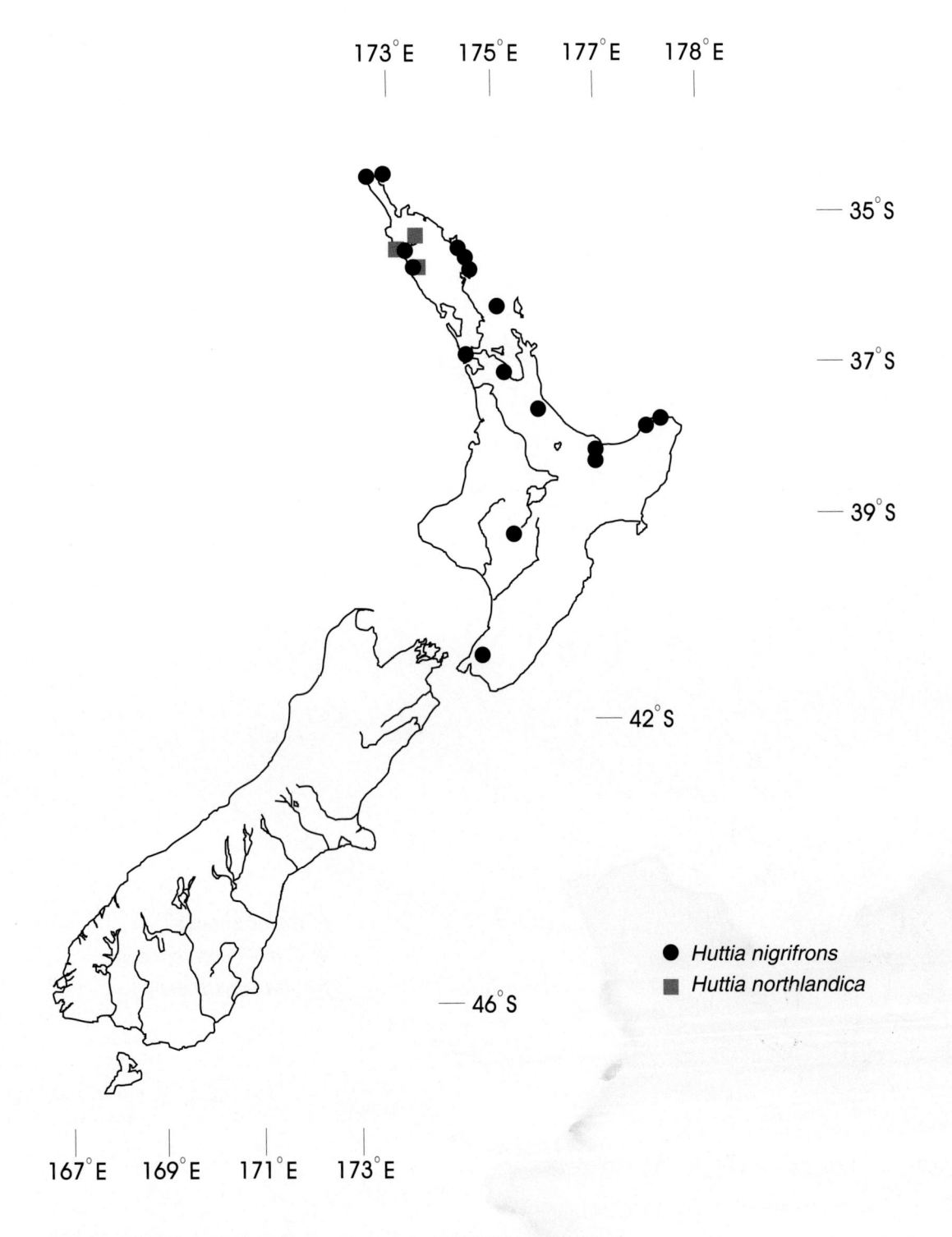

**Map 6** Collection localities of *Huttia* species, mainland New Zealand.

Fauna of New Zealand 40

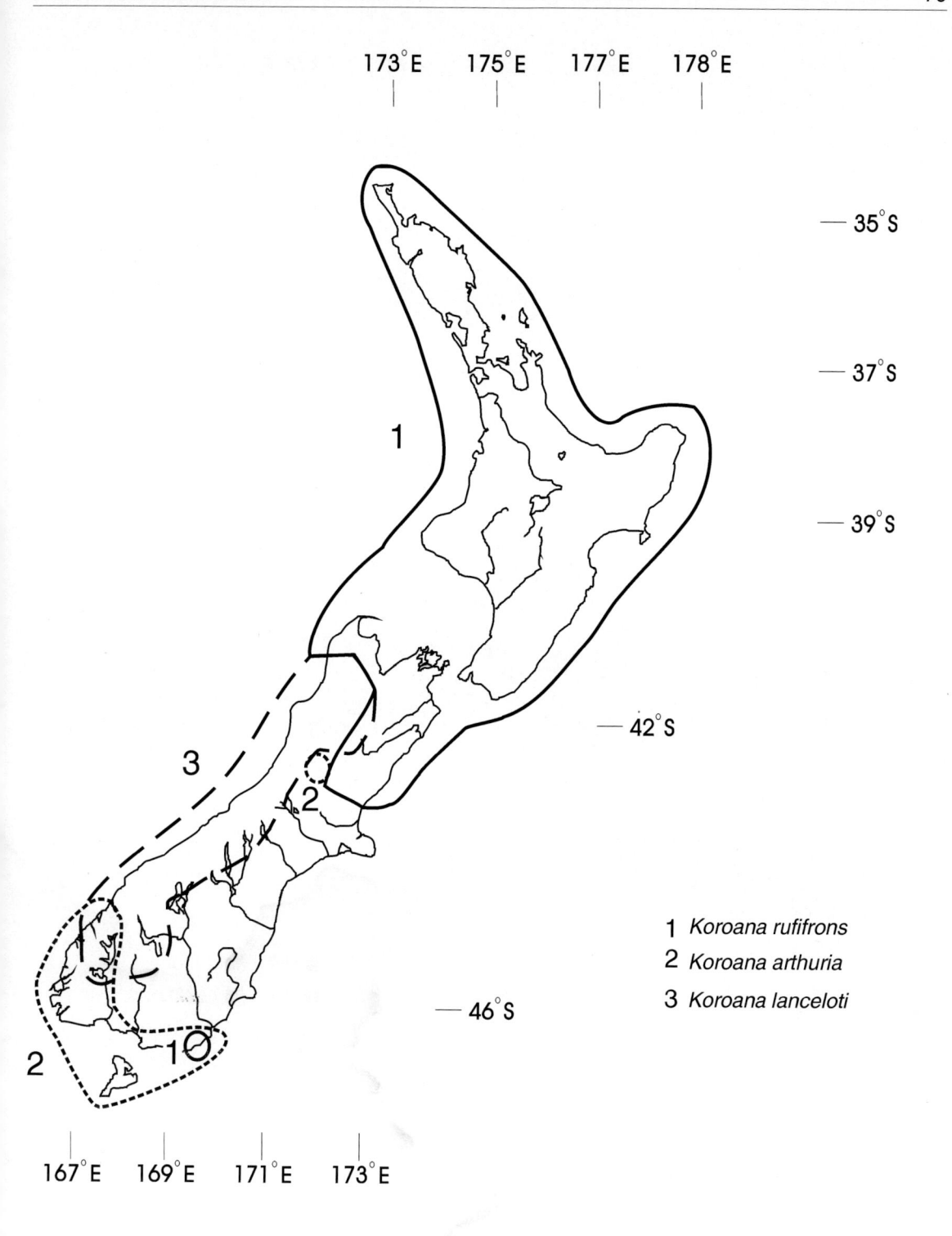

**Map 7** Distribution ranges of *Koroana* species, mainland New Zealand.

1 *Koroana rufifrons*
2 *Koroana arthuria*
3 *Koroana lanceloti*

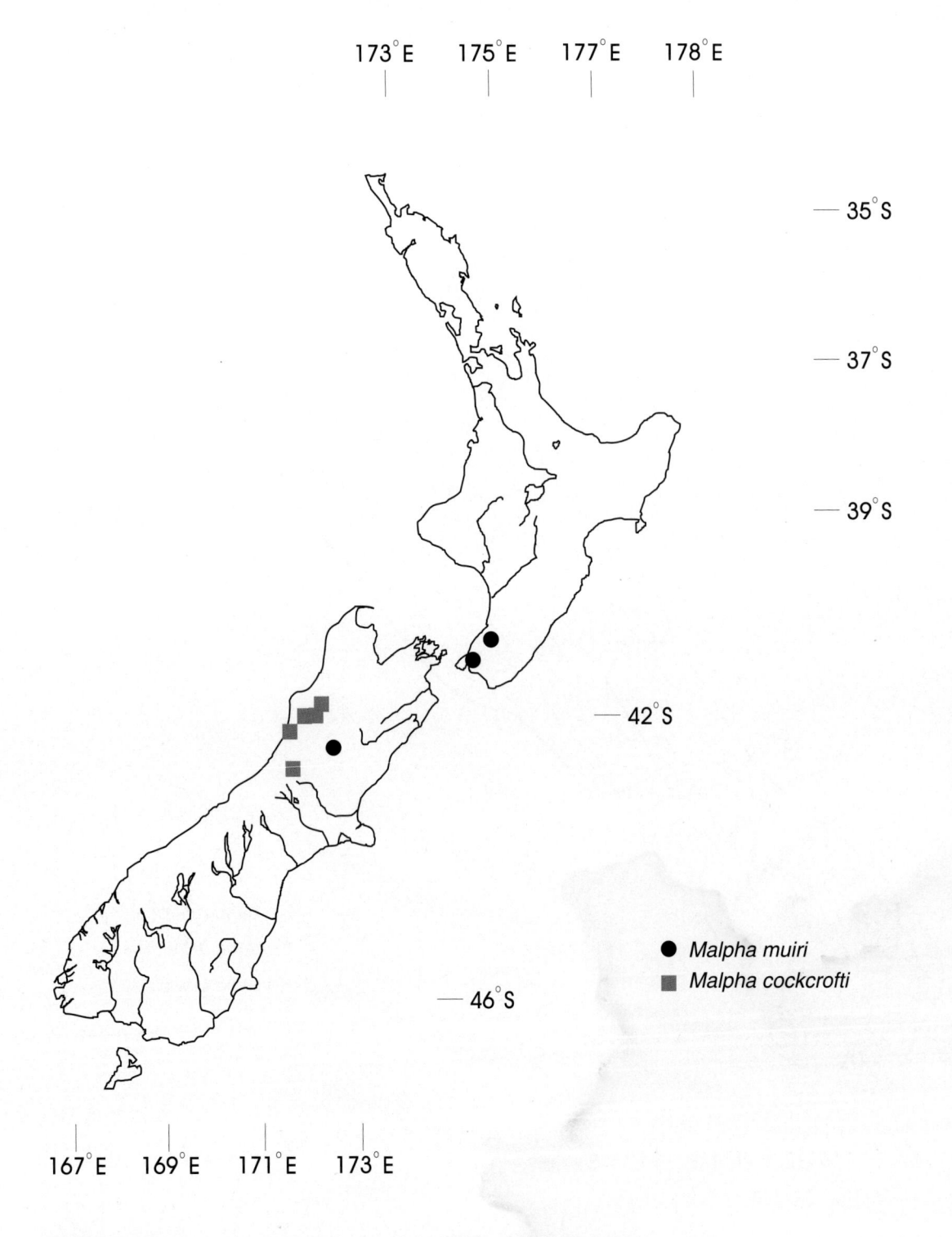

Map 8 Collection localities of *Malpha* species, mainland New Zealand.

Fauna of New Zealand 40

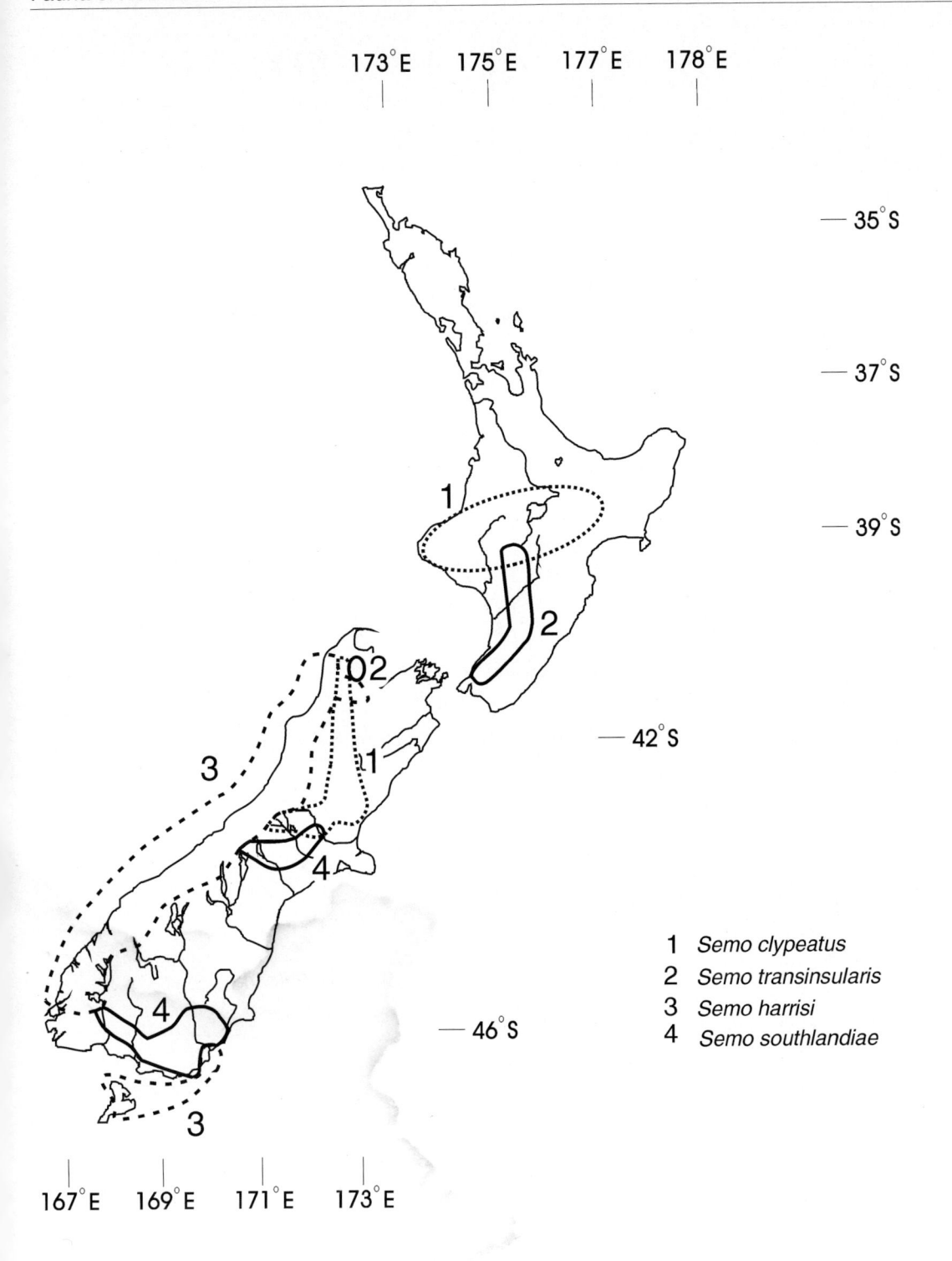

**Map 9** Distribution ranges of *Semo* species, mainland New Zealand.

1 *Semo clypeatus*
2 *Semo transinsularis*
3 *Semo harrisi*
4 *Semo southlandiae*

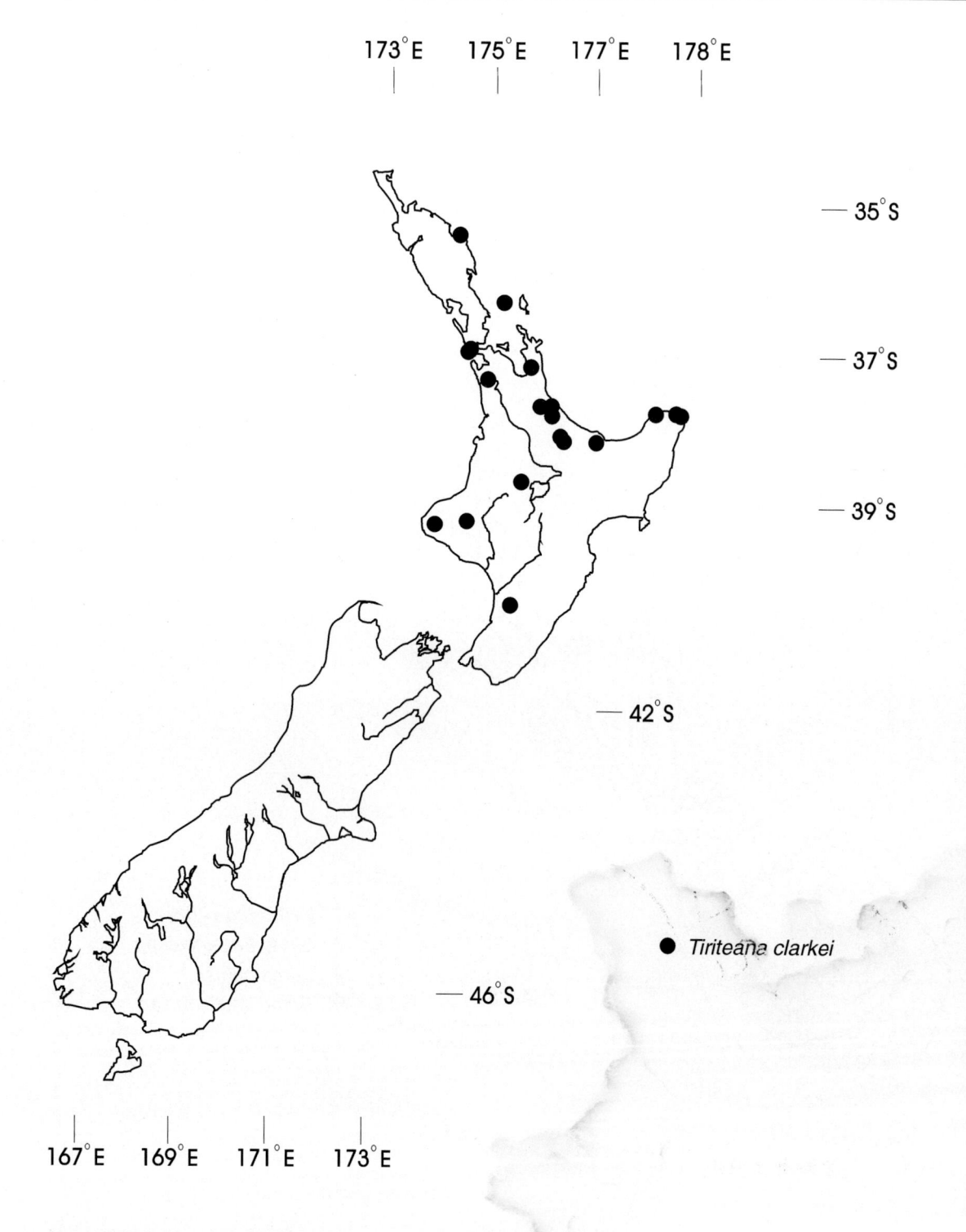

**Map 10** Collection localities of *Tiriteana clarkei*, mainland New Zealand.

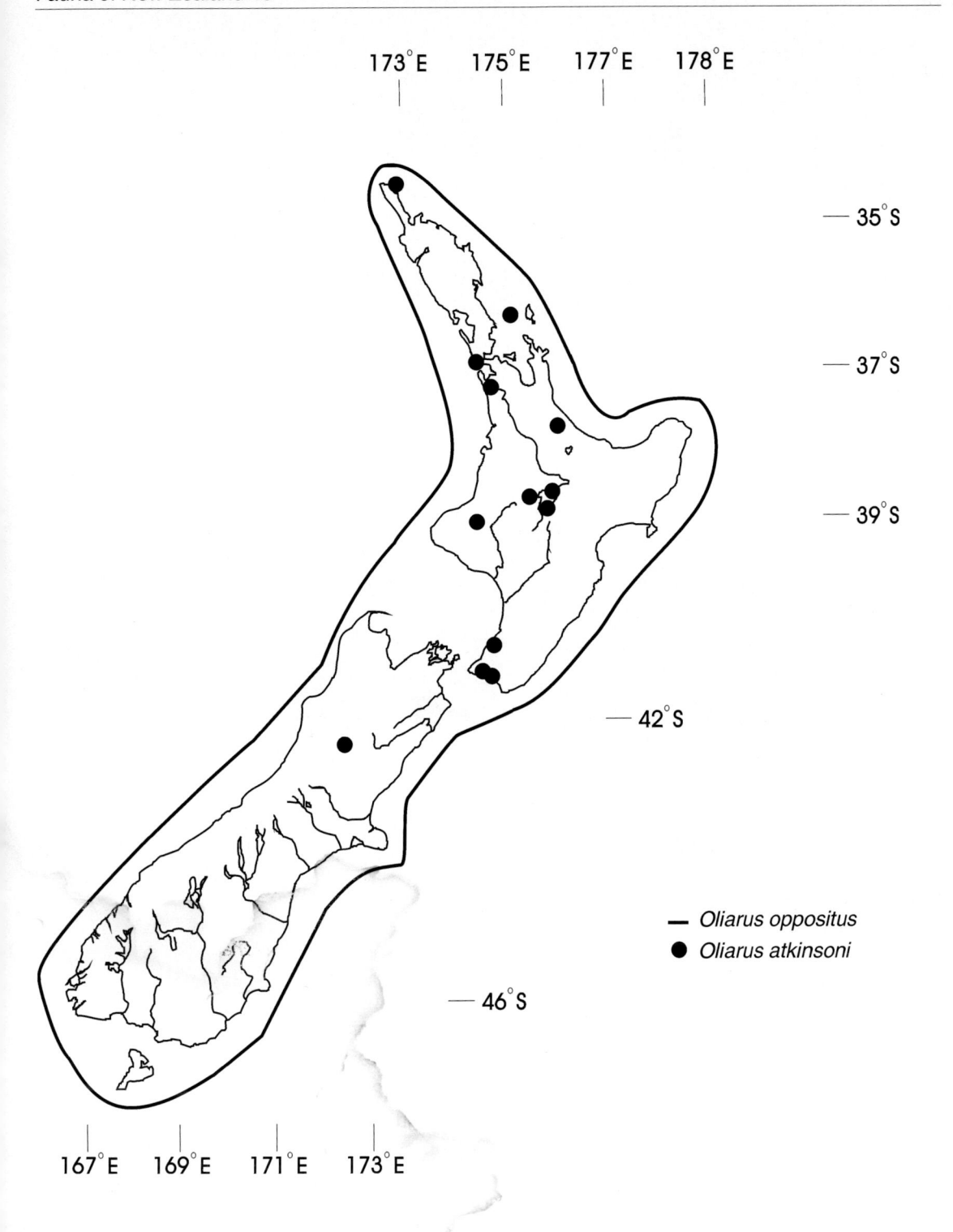

**Map 11** Distribution range of *Oliarus oppositus* and collection localities of *O. atkinsoni*, mainland New Zealand.

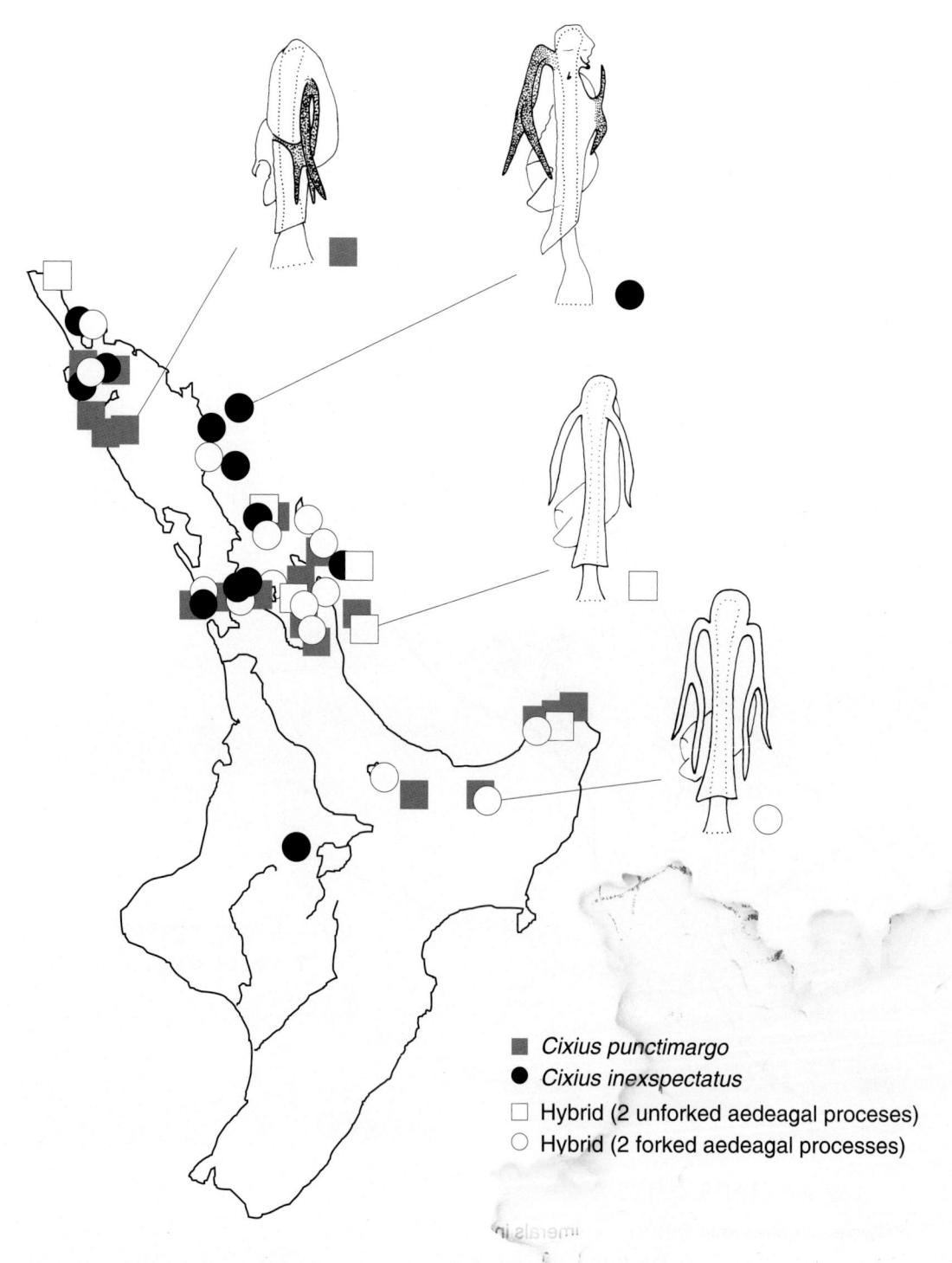

Map 12. Collection localities of *Cixius* species and hybrid populations.

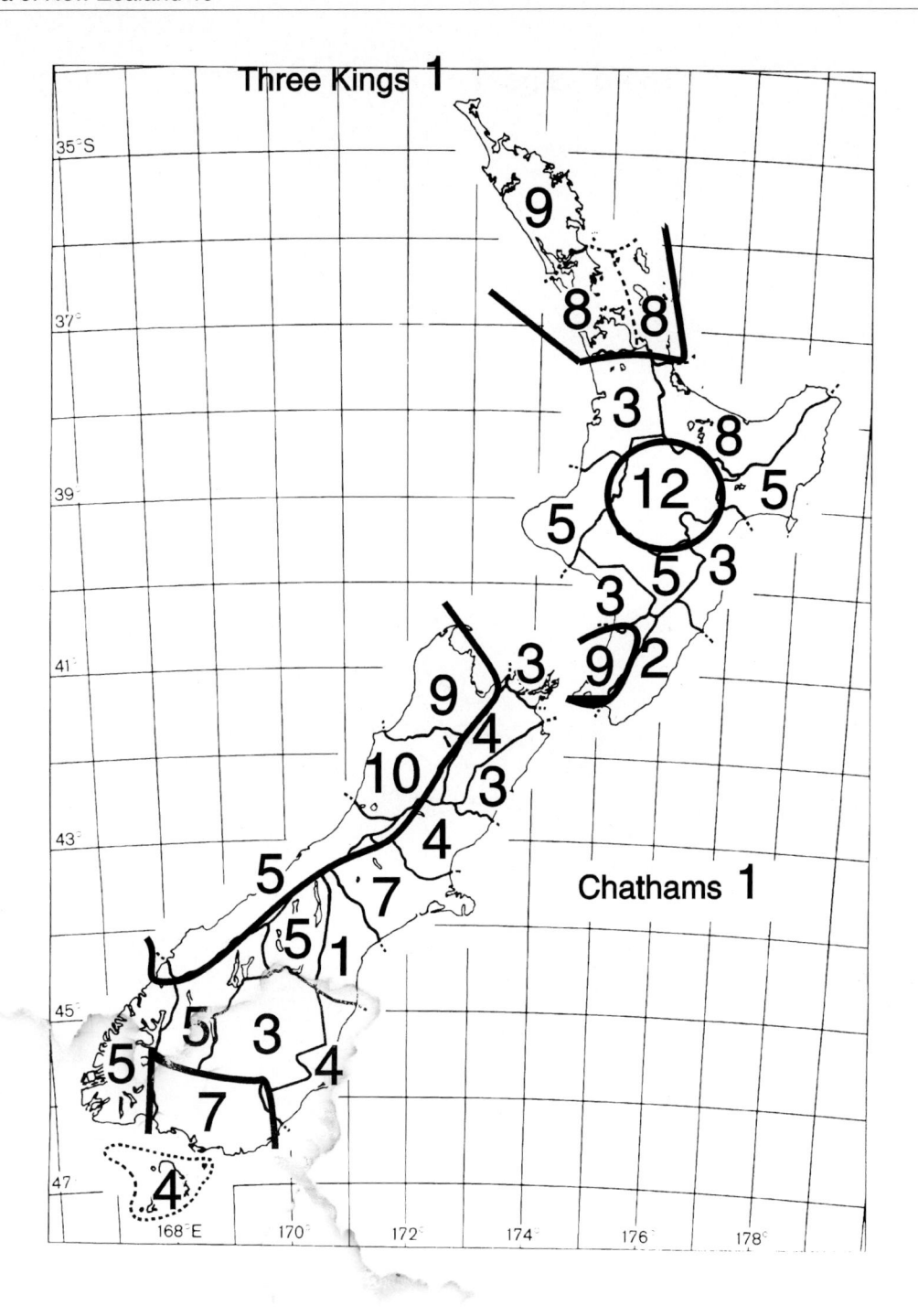

**Map 13. Patterns of taxonomic diversity.** Numbers indicate numbers of species recorded in selected areas of New Zealand. Thick lines denote regions of greatest species diversity. Areas used are those of Crosby et al. (1998).

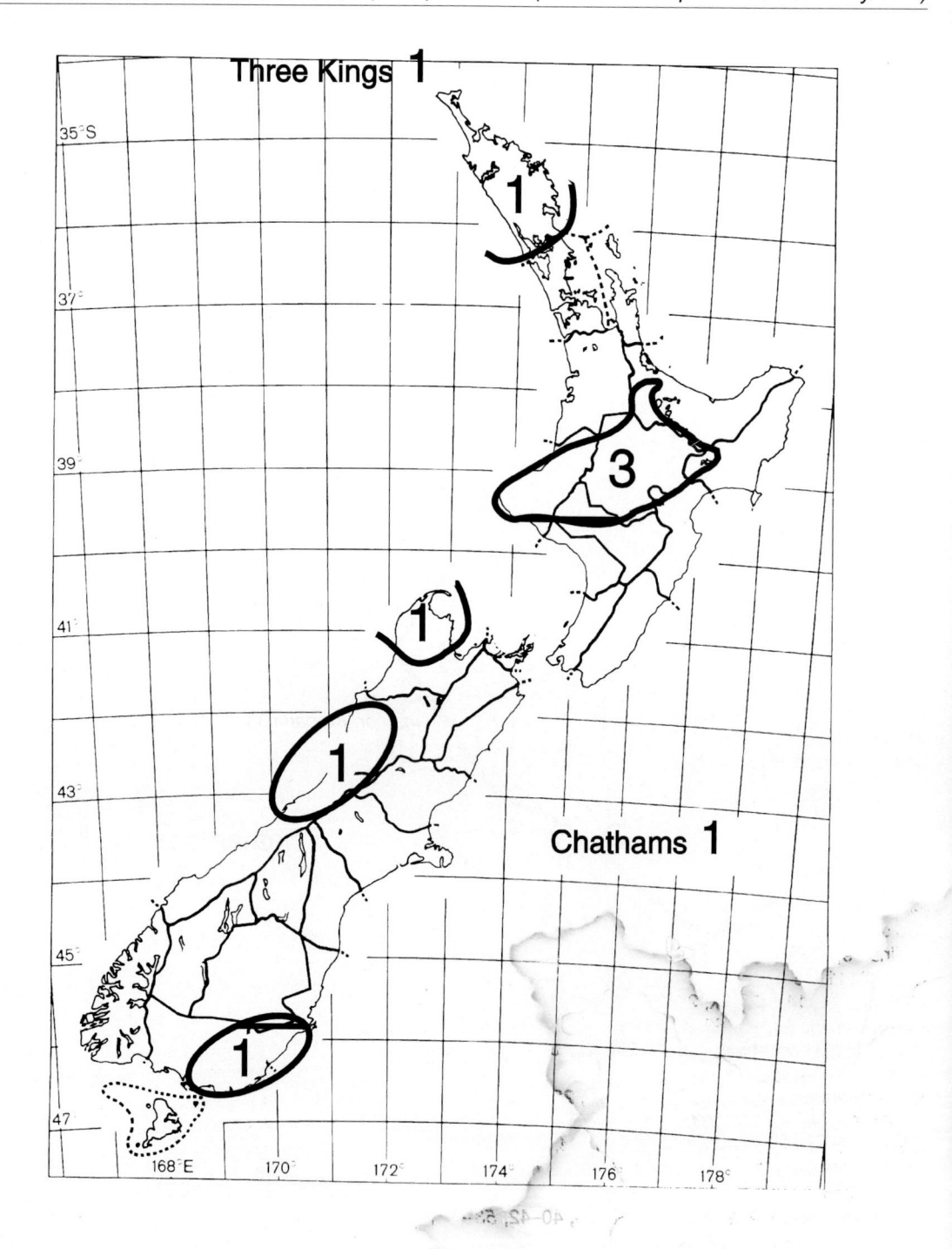

**Map 14. Patterns of species endemism.** Five regions of endemism were identified on mainland New Zealand, and two regions on off-shore islands. Regions of endemism correspond roughly with the regions of greatest species diversity shown on Map 13.

# TAXONOMIC INDEX

Page numbers in bold type denote a description, and in italic type illustrations. A suffixed letter 'k' indicates a key, and 'm' a map.

## A: Invertebrates

*Aka* 5, 11, 12, 14, 15, **18k**, 20, 23
*Aka dunedinensis* 9, 12, 13, **19k**, 21, *61, 64, 76m*
*Aka duniana* 9, 11, 12, 13, **19k**, 20, 21, *61, 64, 76m*
*Aka finitima* 9, 12, 13, 19k, **20**–22, *56, 58, 61, 64, 66, 67, 70, 76m*
*Aka hardyi* 18
*Aka rhodei* 9, 12, 13, 18k, **21**, *61, 64, 76m*
*Aka tasmani* 18
*Aka westlandica* 9, 12, 13, 19k, **22**, 23, *61, 64, 76m*
andrei, *Chathamaka* 6, 9, 12–14, **23**, *58, 61, 64, 66, 67, 70*
arthuria, *Koroana* 9, 12, 13, **31k**–34, *62, 65, 66, 68, 79m*
aspilus, *Cixius* 9, 24, 25
atkinsoni, *Oliarus* 6, 9, 12–15, **44k**, 45, *59, 60, 63, 65, 66, 69, 72, 83m*

*Betacixius* 11
Borystheninae 11
Bothriocerinae 11
Bothriocerini 11

*Chathamaka* 9, 11, 14, 15, 18k, **23**
*Chathamaka andrei* 6, 9, 12–14, **23**, *58, 61, 64, 66, 67, 70*
*Cicada nervosa* 24
Cixiidae 5, 9, 11, 15-17k, 18, 45
Cixiinae 9, 11, 18
Cixiini 9, 11, 18
*Cixius* 5, 11, 14, 15, 18k, **24**, 25k, *60*
*Cixius* hybrids 24, 25, *60, 84m*
*Cixius inexspectatus* 9, 12, 13, **25k**–27, *61, 64, 66, 67, 77m, 84m*
*Cixius punctimargo* 9, 12, 13, 24, 25k, **26**, 27, *58, 61, 64, 66, 70, 77m, 84m*
*Cixius triregius* 9, 12, 13, 25k, **27**, *61, 64, 77m*
clarkei, *Tiriteana* 5, 9, 12, 13, **43**, *59, 63, 64, 66, 69, 72, 82m*
clypeatus, *Semo* 9, 12, 13, 37, 38k, **39**, *53–56, 59, 63, 65, 66, 68, 72, 81m*
cockcrofti, *Malpha* 9, 12, 13, **35k**, 36, *59, 62, 65, 71, 80m*
*Confuga* 11, 14, 15, 17k, **28**
*Confuga persephone* 5, 9, 11–13, **28**, *58, 60, 70*

dunedinensis, *Aka* 9, 12, 13, **19k**, 21, *61, 64, 76m*
duniana, *Aka* 9, 11, 12, 13, **19k**, 20, 21, *61, 64, 76m*
duniana, *Malpha* 19, 20

*Euryphlepsia* 11

finitima, *Aka* 9, 12, 13, 19k, **20**–22, *56, 58, 61, 64, 66, 67, 70, 76m*
finitimus, *Cixius* 18, 20

hardyi, *Aka* 18
harrisi, *Semo* 12, 13, 38k–**40**, 41, *63, 65, 81m*
harrisi, *Huttia* 9, 28, 29, 38, 40
helena, *Koroana* 31, 33
hutchesoni, *Parasemo* 9,12, 13, **37**, *59, 63, 65, 66, 68, 72*
*Huttia* 11, 14, 15, 18k, **28**, 29k, 40
*Huttia nigrifrons* 9, 12, 13, 28, **29k**, 30, *58, 62, 64, 66, 67, 71, 78m*
*Huttia northlandica* 9, 12, 13, 29k, **30**, *62, 64, 71, 78m*

inexspectatus, *Cixius* 9, 12, 13, **25k**–27, *61, 64, 66, 67, 77m, 84m*
interior, *Koroana* 33,
interior, *Cixius* 9, 26, 27, 30–33
iris, *Malpha* 9, 36

kermadecensis, *Cixius* 9, 24
*Koroana* 9, 11–13, 15, 18k, **30**, 31k
*Koroana arthuria* 9, 12, 13, **31k**–34, *62, 65, 66, 68, 79m*
*Koroana* hybrids 31, 33
*Koroana lanceloti* 9, 12, 13, 31k, **32**, 33, *62, 65, 79m*
*Koroana rufifrons* 9, 11–13, 31k–**33**, 34, *58, 62, 65, 71, 79m*
*Kuvera* 11

lanceloti, *Koroana* 9, 12, 13, 31k, **32**, 33, *62, 65, 79m*

*Malpha* 12, 14, 15, 17, 20, **35k**
*Malpha cockcrofti* 9, 12, 13, **35k**, 36, *59, 62, 65, 71, 80m*
*Malpha muiri* 9, 12, 13, 35k, **36**, *59, 62, 65, 66, 68, 80m*
marginalis, *Cixius* 45
muiri, *Malpha* 9, 12, 13, 35k, **36**, *59, 62, 65, 66, 68, 80m*

*nervosa, Cicada* 24
**nigrifrons, Huttia** 9, 12, 13, 28, **29k**, 30, *58, 62, 64, 66, 67, 71, 78m*
**northlandica, Huttia** 9, 12, 13, 29k, **30**, *62, 64, 71, 78m*

Oecleini 9, 11, 43
***Oliarus*** 5, 10, 11, 14, 17k, **44k**
***Oliarus atkinsoni*** 6, 9, 12–15, **44k**, 45, *59, 60, 63, 65, 66, 69, 72, 83m*
***Oliarus oppositus*** 9,12–15, 32, 44k, **45**, *59, 60, 65, 83m*
*oppositus, Cixius* 45
**oppositus, Oliarus** 9,12–15, 32, 44k, **45**, *59, 60, 65, 83m*

***Parasemo*** 9, 11, 14, 15, 17k, **37**
***Parasemo hutchesoni*** 9,12, 13, **37**, *59, 63, 65, 66, 68, 72*
Pentastirini 9, 11, 44
**persephone, Confuga** 5, 9, 11–13, **28**, *58, 60, 70*
**punctimargo, Cixius** 9, 12, 13, 24, 25k, **26**, 27, *58, 61, 64, 66, 70, 77m, 84m*

**rhodei, Aka** 9, 12, 13, 18k, **21**, *61, 64, 76m*
*rufifrons, Cixius* 9, 30, 31, 33
**rufifrons, Koroana** 9, 11–13, 31k–**33**, 34, *58, 62, 65, 71, 79m*

***Semo*** 5, 9, 11–15, 17k, 29, 30, **37**, 38, 38k, *57*
***Semo clypeatus*** 9, 12, 13, 37, 38k, **39**, 40–42, *53–56, 59, 63, 65, 66, 68, 72, 81m*
***Semo harrisi*** 12, 13, 38k–**40**, 41, *63, 65, 81m*
***Semo southlandiae*** 9, 12, 13, 38k, 40, **41**, *63, 65, 81m*
***Semo transinsularis*** 12, 13, 38k–**41**, 42, *63, 65, 81m*
Semonini 11
**southlandiae, Semo** 9, 12, 13, 38k, 40, **41**, *63, 65, 81m*

**tasmani, Aka** 18
***Tiriteana*** 11, 14, 15, 17k, **43**
***Tiriteana clarkei*** 5, 9, 12, 13, **43**, *59, 63, 64, 66, 69, 72, 82m*
**transinsularis, Semo** 12, 13, 38k–**41**, 42, *63, 65, 81m*
**triregius, Cixius** 9, 12, 13, 25k, **27**, *61, 64, 77m*

*walkeri, Cixius* 44
*westlandiae, Semo* 9, 29, 38, 40
**westlandica, Aka** 9, 12, 13, 19k, **22**, 23, *61, 64, 76m*

## B: Plants

*Agathis australis* 6, 15, 52
AGAVACEAE 13, 14
ARALIACEAE 52
ARAUCARIACEAE 52
ARECACEAE 13
*Aristotelia fruticosa* 33, 52
*Ascarina lucida* 23, 52
ASTERACEAE 13, 52

*Beilschmiedia* 15
*Beilschmiedia tarairi* 43, 52
BLECHNACEAE 52
*Blechnum* 6, 15, 23, 52
*Blechnum capense* 23, 52
*Brachyglottis* 15
*Brachyglottis buchananii* 32, 52
*Brachyglottis huntii* 24, 52

*Carmichaelia* 33, 52
*Carpodetus serratus* 15, 23, 43, 52
*Cassinia* 15, 32, 33, 41, 52
*Cassinia vauvilliersii* 39, 52
*Celmisia* 15, 36
*Chionochloa* 42, 52
CHLORANTHACEAE 52
*Coprosma* 6, 15, 21, 22, 23, 26, 33, 39, 41, 43, 52
*Coprosma chathamica* 24, 52
*Coprosma parviflora* 32, 52
*Coprosma propinqua* 40, 52
*Coprosma rhamnoides* 43, 52
*Coriaria arborea* 35, 52
CORIARIACEAE 52
CUNONIACEAE 52
CYPERACEAE 15, 52

*Dacrydium cupressinum* 6, 15, 29, 52
*Dracophyllum* 6, 15, 24, 40, 41, 52
*Dracophyllum longifolium* 39, 40, 52
*Dracophyllum traversii* 40
*Dracophyllum subulatum* 22, 52

ELAEOCARPACEAE 52
EPACRIDACEAE 52
ESCALLONIACEAE 52

FABACEAE 52
FAGACEAE 52
*Fuchsia* 35, 52

GRAMINEAE 15

*Halocarpus* 15
*Halocarpus biformis* 6, 39, 52
*Halocarpus kirkii* 15, 29, 52
*Hebe* 15, 32–35, 39, 40, 52
*Hebe divaricata* 34, 52
*Hebe odora* 32, 41, 52
*Hebe parviflora* 34, 52
*Hebe rakaiensis* 42, 52
*Hebe salicifolia* 33, 52
*Hebe stricta* 34, 39, 52
*Hebe subalpina* 33, 52
*Hoheria* 35, 52

JUNCACEAE 15

LAURACEAE 52

MALVACEAE 52
*Melicytus* 6, 15, 23, 26, 35, 52
*Melicytus chathamicus* 24, 52
*Melicytus ramiflorus* 34, 52
*Metrosideros* 32, 35, 52
MYOPORACEAE 52
*Myoporum* 24, 52
*Myoporum laetum* 27
MYRTACEAE 52

*Nothofagus* 6, 12, 15, 18, 19, 20, 23, 35, 36, 41, 42, 52
*Nothofagus fusca* 21, 39, 52
*Nothofagus menziesii* 21

*Olearia* 15, 33, 36, 39, 52
*Olearia avicenniaefolia* 32, 33, 52
*Olearia colensoi* 36, 52
*Olearia ilicifolia* 40, 52
*Olearia lacunosa* 33, 42, 52
*Olearia moschata* 33, 52
ONAGRACEAE 52

PHORMIACEAE 14, 52
*Phormium* 6, 14, 15, 45, 52
PINACEAE 13
PITTOSPORACEAE 52
*Pittosporum* 35, 52
POACEAE 13, 14, 15, 52
PODOCARPACEAE 52
*Podocarpus ferrugineus* 15
*Podocarpus totara* 6
*Prumnopitys ferruginea* 29, 52
*Pseudopanax* 15, 22, 26, 52
*Pseudopanax crassifolius* 23, 52
*Pseudopanax simplex* (= *Neopanax simplex*) 23, 52
*Pseudowintera* 35, 52

RUBIACEAE 52

*Schefflera* 15
*Schefflera digitata* 23, 52
SCROPHULARIACEAE 52
*Senecio* 15, 36
*Senecio eleagnifolius* 39, 52
SOLANACEAE 52
*Solanum aviculare* var. *albiflorum* 27, 52

*Uncinia* 39, 52

VIOLACEAE 52

*Weinmannia* 35, 52
*Weinmannia racemosa* 23, 52
WINTERACEAE 52

*Xeronema* 21, 52

## TITLES IN PRINT / PUNA TAITARA TAA

1 **Terebrantia** (Insecta: Thysanoptera) ° *Laurence A. Mound & Annette K. Walker*
   ISBN 0-477-06687-9 • 23 Dec 1982 • 120 pp. ............................................................. $29.95

2 **Osoriinae** (Insecta: Coleoptera: Staphylinidae) • *H. Pauline McColl*
   ISBN 0-477-06688-7 • 23 Dec 1982 • 96 pp. ............................................................. $18.60

3 **Anthribidae** (Insecta: Coleoptera) • *B.A. Holloway*
   ISBN 0-477-06703-4 • 23 Dec 1982 • 272 pp. ............................................................ $41.00

4 **Eriophyoidea except Eriophyinae** (Arachnida: Acari) • *D.C.M. Manson*
   ISBN 0-477-06745-X • 12 Nov 1984 • 144 pp. ............................................................ $29.95

5 **Eriophyinae** (Arachnida: Acari: Eriophyoidea) • *D.C.M. Manson*
   ISBN 0-477-06746-8 • 14 Nov 1984 • 128 pp. ............................................................ $29.95

6 **Hydraenidae** (Insecta: Coleoptera) • *R.G. Ordish*
   ISBN 0-477-06747-6 • 12 Nov 1984 • 64 pp. ............................................................. $18.60

7 **Cryptostigmata** (Arachnida: Acari) – a concise review • *M. Luxton*
   ISBN 0-477-06762-X • 8 Dec 1985 • 112 pp. ............................................................. $29.95

8 **Calliphoridae** (Insecta: Diptera) • *James P. Dear*
   ISBN 0-477-06764-6 • 24 Feb 1986 • 88 pp. ............................................................. $18.60

9 **Protura** (Insecta) • *S.L. Tuxen*
   ISBN 0-477-06765-4 • 24 Feb 1986 • 52 pp. ............................................................. $18.60

10 **Tubulifera** (Insecta: Thysanoptera) • *Laurence A. Mound & Annette K. Walker*
   ISBN 0-477-06784-0 • 22 Sep 1986 • 144 pp. ............................................................ $34.65

11 **Pseudococcidae** (Insecta: Hemiptera) • *J.M. Cox*
   ISBN 0-477-06791-3 • 7 Apr 1987 • 232 pp. ............................................................. $49.95

12 **Pompilidae** (Insecta: Hymenoptera) • *A.C. Harris*
   ISBN 0-477-02501-3 • 13 Nov 1987 • 160 pp. ............................................................ $39.95

13 **Encyrtidae** (Insecta: Hymenoptera) • *J.S. Noyes*
   ISBN 0-477-02517-X • 9 May 1988 • 192 pp. ............................................................ $44.95

14 **Lepidoptera** – annotated catalogue, and keys to family-group taxa
   *J. S. Dugdale* • ISBN 0-477-02518-8 • 23 Sep 1988 • 264 pp. ........................................ $49.95

15 **Ambositrinae** (Insecta: Hymenoptera: Diapriidae) • *I.D. Naumann*
   ISBN 0-477-02535-8 • 30 Dec 1988 • 168 pp. ........................................................... $39.95

16 **Nepticulidae** (Insecta: Lepidoptera) • *Hans Donner & Christopher Wilkinson*
   ISBN 0-477-02538-2 • 28 Apr 1989 • 92 pp. ............................................................. $22.95

17 **Mymaridae** (Insecta: Hymenoptera) – introduction, and review of genera
   *J.S. Noyes & E.W. Valentine* • ISBN 0-477-02542-0 • 28 Apr 1989 • 100 pp. .................. $24.95

18 **Chalcidoidea** (Insecta: Hymenoptera) – introduction, and review of genera in smaller families
   *J.S. Noyes & E.W. Valentine* • ISBN 0-477-02545-5 • 2 Aug 1989 • 96 pp. ...................... $24.95

19 **Mantodea** (Insecta), with a review of aspects of functional morphology
   and biology • *G.W. Ramsay* • ISBN 0-477-02581-1 • 13 Jun 1990 • 96 pp. ....................... $24.95

20 **Bibionidae** (Insecta: Diptera) • *Roy A. Harrison*
   ISBN 0-477-02595-1 • 13 Nov 1990 • 28 pp. ............................................................. $14.95

21 **Margarodidae** (Insecta: Hemiptera) • *C.F. Morales*
ISBN 0-477-02607-9 • 27 May 1991 • 124 pp. .................................................................. $34.95

22 **Notonemouridae** (Insecta: Plecoptera) • *I.D. McLellan*
ISBN 0-477-02518-8 • 27 May 1991 • 64 pp. .................................................................... $24.95

23 **Sciapodinae, Medeterinae** (Insecta: Diptera) with a generic review of the
Dolichopodidae • *D.J. Bickel* • ISBN 0-477-02627-3 • 13 Jan 1992 • 74 pp. ..................... $27.95

24 **Therevidae** (Insecta: Diptera) • *L. Lyneborg*
ISBN 0-477-02632-X • 4 Mar 1992 • 140 pp. .................................................................... $34.95

25 **Cercopidae** (Insecta: Homoptera) • *K.G.A. Hamilton & C.F. Morales*
ISBN 0-477-02636-2 • 25 May 1992 • 40 pp. .................................................................... $17.95

26 **Tenebrionidae** (Insecta: Coleoptera): catalogue of types and keys to taxa
*J.C. Watt* • ISBN 0-477-02639-7 • 13 Jul 1992 • 70 pp. .................................................... $27.95

27 **Antarctoperlinae** (Insecta: Plecoptera) • *I.D. McLellan*
ISBN 0-477-01644-8 • 18 Feb 1993 • 70 pp. ...................................................................... $27.95

28 **Larvae of Curculionoidea** (Insecta: Coleoptera): a systematic overview
*Brenda M. May* • ISBN 0-478-04505-0 • 14 Jun 1993 • 226 pp. ....................................... $55.00

29 **Cryptorhynchinae** (Insecta: Coleoptera: Curculionidae)
*C.H.C. Lyal* • ISBN 0-478-04518-2 • 2 Dec 1993 • 308 pp. .............................................. $65.00

30 **Hepialidae** (Insecta: Lepidoptera) • *J.S. Dugdale*
ISBN 0-478-04524-7 • 1 Mar 1994 • 164 pp. ...................................................................... $42.50

31 **Talitridae** (Crustacea: Amphipoda) • *K.W. Duncan*
ISBN 0-478-04533-6 • 7 Oct 1994 • 128 pp. ...................................................................... $36.00

32 **Sphecidae** (Insecta: Hymenoptera) • *A.C. Harris*
ISBN 0-478-04534-4 • 7 Oct 1994 • 112 pp. ...................................................................... $33.50

33 **Moranilini** (Insecta: Hymenoptera) • *J.A. Berry*
ISBN 0-478-04538-7 • 8 May 1995 • 82 pp. ...................................................................... $29.95

34 **Anthicidae** (Insecta: Coleoptera) • *F.G. Werner & D.S. Chandler*
ISBN 0-478-04547-6 • 21 Jun 1995 • 64 pp. ...................................................................... $26.50

35 **Cydnidae, Acanthosomatidae, and Pentatomidae** (Insecta: Heteroptera):
systematics, geographical distribution, and bioecology • *M.-C. Larivière*
ISBN 0-478-09301-2 • 23 Nov 1995 • 112 pp. .................................................................... $42.50

36 **Leptophlebiidae** (Insecta: Ephemeroptera) • *D.R. Towns & W.L. Peters*
ISBN 0-478-09303-9 • 19 Aug 1996 • 144 pp. .................................................................... $39.50

37 **Coleoptera**: family-group review and keys to identification • *J. Klimaszewski
& J.C. Watt* • ISBN 0-478-09312-8 • 13 Aug 1997 • 199 pp. ............................................. $49.50

38 **Naturalised terrestrial Stylommatophora** (Mollusca: Gastropoda)
*G.M. Barker* • ISBN 0-478-09322-5 • 25 Jan 1999 • 253 pp. ............................................ $72.50

39 **Molytini** (Insecta: Coleoptera: Curculionidae: Molytinae) • *R.C. Craw*
ISBN 0-478-09325-X • 4 Feb 1999 • 68 pp. ........................................................................ $29.50

40 **Cixiidae** (Insecta: Hemiptera: Auchenorrhyncha) • *M.-C..Larivière*
ISBN 0-478-09334-9 • 1999 • 93 pp. ................................................................................. $37.50

Visit the Manaaki Whenua Press Website at http://www.mwpress.co.nz/ for further information, and to gain access to on-line extracts from these publications.

## Taxonomic groups covered in the *Fauna of New Zealand* series

### Insecta

**Coleoptera**
Family-group review and keys to identification (*J. Klimaszewski & J.C. Watt*, FNZ 37, 1997)
Anthribidae (*B.A. Holloway*, FNZ 3, 1982)
Anthicidae (*F.G. Werner & D.S. Chandler*, FNZ 34, 1995)
Curculionidae: Cryptorhynchinae (*C.H.C. Lyal*, FNZ 29, 1993)
Curculionidae: Molytinae: Molytini (*R. C. Craw*, FNZ 39, 1999)
Curculionoidea larvae: a systematic overview (*Brenda M. May*, FNZ 28, 1993)
Hydraenidae (*R.G. Ordish*, FNZ 6, 1984)
Staphylinidae: Osoriinae (*H. Pauline McColl*, FNZ 2, 1982)
Tenebrionidae: catalogue of types and keys to taxa (*J.C. Watt*, FNZ 26, 1992)

**Diptera**
Bibionidae (*Roy A. Harrison*, FNZ 20, 1990)
Calliphoridae (*James P. Dear*, FNZ 8, 1986)
Dolichopodidae: Sciapodinae, Medeterinae with a generic review (*D.J. Bickel*, FNZ 23, 1992)
Therevidae (*L. Lyneborg*, FNZ 24, 1992)

**Ephemeroptera**
Leptophlebiidae (*D.R. Towns & W.L. Peters*, FNZ 36, 1996)

**Hemiptera**
Cercopidae (*K.G.A. Hamilton & C.F. Morales*, FNZ 25, 1992)
Cixiidae (*M.-C. Larivière*, FNZ 40, 1999)
Cydnidae, Acanthosomatidae, and Pentatomidae (*M.-C. Larivière*, FNZ 35, 1995)
Margarodidae (*C.F. Morales*, FNZ 21, 1991)
Pseudococcidae (*J.M. Cox*, FNZ 11, 1987)

**Hymenoptera**
Chalcidoidea: introduction, and review of smaller families (*J.S. Noyes & E.W. Valentine*, FNZ 18, 1989)
Diapriidae: Ambositrinae (*I.D. Naumann*, FNZ 15, 1988)
Encyrtidae (*J.S. Noyes*, FNZ 13, 1988)
Mymaridae (*J.S. Noyes & E.W. Valentine*, FNZ 17, 1989)
Pompilidae (*A.C. Harris*, FNZ 12, 1987)
Pteromalidae: Eunotinae: Moranilini (*J.A. Berry*, FNZ 33, 1995)
Sphecidae (*A.C. Harris*, FNZ 32, 1994)

**Lepidoptera**
Annotated catalogue, and keys to family-group taxa (*J. S. Dugdale*, FNZ 14, 1988)
Hepialidae (*J.S. Dugdale*, FNZ 30, 1994)
Nepticulidae (*Hans Donner & Christopher Wilkinson*, FNZ 16, 1989)

**Mantodea**, with a review of aspects of functional morphology and biology (*G.W. Ramsay*, FNZ 19, 1990)

**Plecoptera**
Antarctoperlinae (*I.D. McLellan*, FNZ 27, 1993)
Notonemouridae (*I.D. McLellan*, FNZ 22, 1991)

**Protura** (*S.L. Tuxen*, FNZ 9, 1986)

**Thysanoptera**
Terebrantia (*Laurence A. Mound & Annette K. Walker*, FNZ 1, 1982)
Tubulifera (*Laurence A. Mound & Annette K. Walker*, FNZ 10, 1986)

### Arachnida

**Acari**
Cryptostigmata – a concise review (*M. Luxton*, FNZ 7, 1985)
Eriophyoidea except Eriophyinae (*D.C.M. Manson*, FNZ 4, 1984)
Eriophyinae (*D.C.M. Manson*, FNZ 5, 1984)

### Crustacea

**Amphipoda**
Talitridae (*K.W. Duncan*, FNZ 31, 1994)

### Mollusca

**Gastropoda**
Naturalised terrestrial Stylommatophora (*G.M. Barker*, FNZ 38, 1999)

## NOTICES

This series of refereed publications has been established to encourage those with expert knowledge to publish concise yet comprehensive accounts of elements in the New Zealand fauna. The series is professional in its conception and presentation, yet every effort is made to provide resources for identification and information that are accessible to the non-specialist.

*Fauna of N.Z.* deals with non-marine invertebrates only, since the vertebrates are well documented, and marine forms are covered by the series *Marine Fauna of N.Z.*

**Contributions** are invited from any person with the requisite specialist skills and resources. Material from the N.Z. Arthropod Collection is available for study.

Contributors should discuss their intentions with a member of the Invertebrate Systematics Advisory Group or with the Series Editor before commencing work; all necessary guidance will be given.

**Subscribers** should address inquiries to *Fauna of N.Z.*, Manaaki Whenua Press, Landcare Research, P.O. Box 40, Lincoln 8152, New Zealand.

Subscription categories: 'A' – standing orders; an invoice will be sent with each new issue, as soon after publication as possible; 'B' – promotional fliers with order forms will be sent from time to time.

Retail prices (see 'Titles in print', page 66) include packaging and surface postage. Subscribers in New Zealand and Australia pay the indicated amount in $NZ; GST is included in the price. Other subscribers pay the listed price in $US, or its equivalent.

Back issues of all numbers are available, and new subscribers wishing to obtain a full set or a selection may request a discount. Booksellers and subscription agents are offered a trade discount of ten percent.

## NGĀ PĀNUI

Kua whakatūria tēnei huinga pukapuka hei whakahauhau i ngā tohunga whai mātauranga kia whakaputa i ngā kōrero poto, engari he whaikiko tonu, e pā ana ki ngā aitanga pepeke o Aotearoa. He tōtika tonu te āhua o ngā tuhituhi, engari ko te tino whāinga, kia mārama te marea ki ngā tohu tautuhi o ia ngārara, o ia ngārara, me te roanga atu o ngā kōrero mō tēnā, mō tēnā.

He titiro whāiti tā tēnei pukapuka ki ngā mea noho whenua, kāore he tuarā; i pēnei ai i te mea kei te mōhio whānuitia ngā mea whai tuarā, ā, ko ngā mea noho moana, koirā te tino kaupapa o te huinga pukapuka *Marine Fauna of N.Z.*

Ka āhei te tangata ki te **whakauru tuhituhinga** mehemea kei a ia ngā tohungatanga me ngā rauemi e tutuki pai ai tana mahi. Heoi anō, e wātea ana te Kohinga Angawaho o Aotearoa hei āta tirotiro mā te tangata mehemea he āwhina kei reira.

Me whāki te kaituhi i ōna whakaaro ki tētahi o te Kāhui Ārahi Whakarōpūtanga Tuarā-Kore, ki te Ētita rānei i mua i te tīmatanga, ā, mā rātou a ia e ārahi mō te wāhi ki tana tuhinga.

Ko te hunga pīrangi **hoko pukapuka**, me tuhi ki *Fauna of N.Z.*, Manaaki Whenua Press, Manaaki Whenua, Pouaka Poutāpeta 40, Lincoln 8152, Aotearoa.

E rua ngā tūmomo kaihoko: "A" – kaihoko tūmau, ka tukua ia pukapuka, ia pukapuka, me te nama, i muri tonu i te tānga; "B" – ka tukua ngā pānui whakatairanga me ngā puka tono i ōna wā anō.

Te utu (tirohia "Titles in print", whārangi 66). Ko te kōpaki me te pane kuini kei roto i te utu. Me utu te hunga e noho ana i Aotearoa me Ahitereiria ki ngā tāra o Aotearoa. Ko ētahi atu me utu te moni kua tohua, ki ngā tāra Merikana, ki te nui o te moni rānei e rite ana.

E toe ana he pukapuka o ngā putanga katoa o mua. Mehemea e hiahia ana koe ki te katoa o ngā pukapuka, ki ētahi rānei, tonoa mai kia whakahekea te utu. Tekau ōrau te heke iho o te utu ki ngā toa hoko pukapuka.